W9-BTP-623

Praise for *War of the Worldviews*

"We physicists are concerned with observations of the physical universe and the mathematical theories that explain them. Others seek enlightenment through a focus on subjective experience. In this book these approaches meet, often throwing off sparks, occasionally agreeing, and always remaining both illuminating and entertaining."

—Jay Marx, Executive Director, Laser Interferometer Gravitational-wave Observatory (LIGO) Laboratory, Caltech

"Most conflicts in life can be traced to worldview differences, and none more so than the worldviews of science and religion. *War of the Worldviews* is the best single volume I've ever read on this vital subject. Deepak Chopra and Leonard Mlodinow well capture the essence of the debate and do so in such an engaging style that you can't stop reading. This book is a game changer in the science-and-religion wars."

—Michael Shermer, publisher of *Skeptic* magazine, monthly columnist for *Scientific American,* adjunct professor at Claremont Graduate University and Chapman University, author of *Why Darwin Matters* and *The Believing Brain*

"A refreshing and more useful approach to the old combat between science and religion. The two authors want the best for humanity, and their zeal is revealed even when they fiercely disagree. The value of this book will only become greater and more appreciated with time."

—Menas Kafatos, Ph.D., Fletcher Jones Endowed Professor in Computational Physics; Dean, Schmid College of Science, Vice Chancellor for Special Projects, Chapman University

"Quantum mechanics demonstrates the reality of particle entanglement. The reality of today's world is that all of our lives are entangled. The dialogue between these two extraordinary writers serves as a source of awe and inspiration to all of us."

—James R. Doty, M.D., Professor of Neurosurgery, Founder and Director, Center for Compassion and Altruism Research and Education (CCARE), Stanford Institute of Neuro-innovation and Translational Neuroscience, Stanford University School of Medicine

"Astrophysicist Sir James Jeans wrote: 'The Universe begins to look more like a great thought than like a great machine.' This is the essence of Chopra's view: that a great consciousness—which we share—is the basis of the Universe and all reality. From Mlodinow's perspective it is unimaginable that consciousness could be anything more than brain chemistry at work and certainly not something capable of creating a universe. The book presents a lively and articulate debate on this most important human question: Are we simply complex biological machines destined for oblivion at death, or are we immortal spiritual beings temporarily experiencing reality through physical bodies?"

—BERNARD HAISCH, astrophysicist

"Deepak Chopra and Leonard Mlodinow argue convincingly for their particular worldviews. However, reading this book convinces me they should call a truce: science and spirituality are two sides of a quantum coin."

—STUART HAMEROFF, M.D., Professor, Anesthesiology and Psychology; Director, Center for Consciousness Studies, The University of Arizona, Tucson

"Finally! The beginning of a dialogue in the true spirit of open-ended science that should be inclusive of all phenomena including spirituality. Congratulations to Chopra and Mlodinow for the breakthrough."

—AMIT GOSWAMI, quantum physicist and author of *The Self-Aware Universe* and *How Quantum Activism Can Save Civilization*

"Is consciousness an aspect of nature that had no precursor prior to the appearance of life, or is it a feature of nature that was in some form always present? This question is debated in this lively, informative, and entertaining book coauthored by two skilled writers. Chopra argues for the pervasiveness of consciousness, while Mlodinow argues for emergence of everything from the purely physical, in the absence of adequate scientific evidence to the contrary. This book is a good read even if, and particularly if, you already have a fixed opinion on the matter."

—DR. HENRY P. STAPP, physicist, Lawrence Berkeley National Laboratory, University of California, Berkeley, and author of *Mind, Matter, and Quantum Mechanics* and *Mindful Universe: Quantum Mechanics and the Participating Observer*

"Deepak Chopra and Leonard Mlodinow have opened the discussion on the fundamental physics of the spirit."

—Juliana (Brooks) Mortenson, M.D., Founder, General Resonance

"Ours is a time of unprecedented change and complexity. Never before have so many worldviews, belief systems, and ways of engaging reality converged. Such a moment of contact has many consequences. On the one hand, there are abundant instances of conflict and intolerance, as people fail to see other points of view. On the other hand, the situation can lead to the creative emergence of new and more sustainable ways of being together in our otherwise fragmented world. Such is the promise of this thoughtful and provocative book. As Chopra and Mlodinow, two masters in their respective fields, come together to consider the challenges of merging science and spirituality, they offer an essential guidebook for shaping the future of our shared humanity."

—Marilyn Schlitz, Ph.D., President and CEO, Institute of Noetic Sciences

"In this latest skirmish of the age-old War of the Worldviews, we find a spirited defense of both science and spirituality. The authors are masters of their domains, and their debate makes it crystal clear that the battle will not be settled any time soon. Reading this book may make your brain hurt, but it is an experience that is fascinating, exasperating, and definitely worthwhile."

—Dean Radin, Ph.D., Co-Editor-in-Chief, *Explore: The Journal of Science and Healing;* Adjunct Professor, Department of Psychology, Sonoma State University; Senior Scientist, Institute of Noetic Sciences

"In *War of the Worldviews*, Chopra and Mlodinow prove to be eloquent proponents for their respective points of view. Though it is clear they remain far apart on many issues, the mere act of these two acclaimed thinkers addressing them together provides hope that the divide between science and spirituality can be narrowed."

—Jim B. Tucker, M.D., Division of Perceptual Studies, Department of Psychiatry and Neurobehavioral Sciences, University of Virginia Health System

"A tension exists between the way that we think about the laws of physics and our own subjective experience. Chopra and Mlodinow ponder both perspectives in their lively debate, leaving the reader enriched to see the world with a new depth. *War of the Worldviews* offers clear choices for these rapidly changing times."

—Jeff Tollaksen, Director, Center for Quantum Studies, Head of Physics Faculty, Schmid College of Science, Chapman University

"As a brilliant scientist and mathematician, Leonard Mlodinow believes that physics can account for the creation of the universe through the laws of nature, without the participation of a deity. To Deepak Chopra, the truth exists in consciousness. The time has come for humanity to open its mind to all levels of reality."

—Lothar Schäfer, Distinguished Professor of Chemistry and Biochemistry, University of Arkansas

"Deepak Chopra did an excellent job explaining why the all-embracing quantum field suggests a dynamic, alive cosmos. This is an interesting and provocative book that will be read and talked about for a long time to come."

—Hans Peter Duerr, Director Emeritus, Max Planck Institute for Physics and Astrophysics

"*War of the Worldviews* offers a fascinating and detailed debate focusing on how the spiritual and the scientific approaches to understanding reality often clash. Physician Deepak Chopra and physicist Leonard Mlodinow provide a rich set of reflections and easy-to-understand introductions to the various topics, from the nature of mind and consciousness to God and the brain. Diving into the conceptual friction and heated emotional tension of this important and passionate conversation between two leaders in these fields inspires us to weave a tapestry of our own, blending the hard-won insights from an empirical approach to reality with the important journey to make a life of meaning and interconnection in our daily lives."

—Daniel J. Siegel, M.D., author of *Mindsight: The New Science of Personal Transformation,* Clinical Professor, UCLA School of Medicine, Executive Director, Mindsight Institute

WAR OF THE WORLDVIEWS

Also by Leonard Mlodinow

Euclid's Widow: The Story of Geometry from Parallel Lines to Hyperspace

Feynman's Rainbow: A Search for Beauty in Physics and in Life

A Briefer History of Time (with Stephen Hawking)

The Drunkard's Walk: How Randomness Rules Our Lives

The Grand Design (with Stephen Hawking)

FOR CHILDREN

The Last Dinosaur (with Matt Costello)

Titanic Cat (with Matt Costello)

Also by Deepak Chopra

Consciousness in the Universe: Quantum Physics, Evolution, Brain and Mind (with Stuart Hameroff and Sir Roger Penrose)

Creating Health

Return of the Rishi

Quantum Healing

Perfect Health

Unconditional Life

Ageless Body, Timeless Mind

Creating Affluence

The Seven Spiritual Laws of Success

The Seven Spiritual Laws of Success: Pocket Guide

The Way of the Wizard

The Return of Merlin

Boundless Energy

Journey into Healing

The Path to Love

Perfect Weight

Restful Sleep

Raid on the Inarticulate

The Seven Spiritual Laws for Parents

Perfect Digestion

Overcoming Addictions

The Love Poems of Rumi (edited by Deepak Chopra; translated by Deepak Chopra and Fereydoun Kia)

Healing the Heart

Everyday Immortality

On the Shores of Eternity

The Daughters of Joy

The Lords of Light

The Angel Is Near

How to Know God

The Deeper Wound

The Chopra Center Herbal Handbook (with coauthor David Simon)

Grow Younger, Live Longer (with coauthor David Simon)

The Soul in Love

Soulmate

Golf for Enlightenment

The Spontaneous Fulfillment of Desire

Manifesting Good Luck

The Chopra Center Cookbook (coauthored by David Simon and Leanne Backer)

The Book of Secrets

Fire in the Heart

Peace Is the Way

The Seven Spiritual Laws of Yoga (with coauthor David Simon)

Magical Beginnings, Enchanted Lives (coauthored by David Simon and Vicki Abrams)

Teens Ask Deepak

Ask the Kabala (with coauthor Michael Zappolin)

Power, Freedom, and Grace

Life After Death

Kama Sutra

Buddha

The Essential How to Know God

The Essential Spontaneous Fulfillment of Desire

The Essential Ageless Body, Timeless Mind

The Third Jesus

Why Is God Laughing?

Jesus

Reinventing the Body, Resurrecting the Soul

The Ultimate Happiness Prescription

The Shadow Effect (with coauthors Debbie Ford and Marianne Williamson)

Muhammad

The Soul of Leadership

Walking Wisdom (contributor; authored by Gotham Chopra)

The Seven Spiritual Laws of Superheroes (with coauthor Gotham Chopra)

FOR CHILDREN

On My Way to a Happy Life (with Kristina Tracy,
illustrated by Rosemary Woods)

You with the Stars in Your Eyes (illustrated by Dave Zaboski)

DEEPAK CHOPRA

WAR OF THE WORLDVIEWS

SCIENCE vs. SPIRITUALITY

LEONARD MLODINOW

HARMONY BOOKS NEW YORK

Published in the United States by Harmony Books, an imprint of the Crown Publishing Group, a division of Random House, Inc., New York.
www.crownpublishing.com

Harmony Books is a registered trademark and the Harmony Books colophon is a trademark of Random House, Inc.

Library of Congress Cataloging-in-Publication Data
Chopra, Deepak.
War of the worldviews : science vs. spirituality / by Deepak Chopra and Leonard Mlodinow.—1st ed.
p. cm.
Includes index.
1. Religion and science. I. Mlodinow, Leonard, 1954– II. Title.
BL240.3.C46 2011
201'.65—dc22 2011010591

ISBN 978-0-307-88688-0
eISBN 978-0-307-88690-3

Printed in the United States of America

Book design: Lauren Dong
Jacket design: Daniel Rembert
Jacket photographs: (burst) Dimitri Vervitsioris/Getty Images, (sun) Stephen Coburn/Bigstock.com

1 3 5 7 9 10 8 6 4 2

First Edition

To all the sages and scientists
who have expanded the human mind

Contents

Foreword by Deepak and Leonard *xvii*

Part One: THE WAR

1. PERSPECTIVES 3
   The Spiritual Perspective: Deepak *4*
   The Scientific Perspective: Leonard *11*

Part Two: COSMOS

2. HOW DID THE UNIVERSE EMERGE? 23
   Leonard *24*
   Deepak *32*

3. IS THE UNIVERSE CONSCIOUS? 39
   Deepak *40*
   Leonard *46*

4. IS THE UNIVERSE EVOLVING? 51
   Deepak *52*
   Leonard *57*

5. WHAT IS THE NATURE OF TIME? 65
   Leonard *66*
   Deepak *72*

6. Is the Universe Alive? 79
Deepak 80
Leonard 85

Part Three: LIFE

7. What Is Life? 93
Leonard 94
Deepak 101

8. Is There Design in the Universe? 107
Leonard 108
Deepak 117

9. What Makes Us Human? 121
Deepak 122
Leonard 126

10. How Do Genes Work? 135
Leonard 136
Deepak 144

11. Did Darwin Go Wrong? 151
Deepak 152
Leonard 161

Part Four: MIND AND BRAIN

12. What Is the Connection Between Mind and Brain? 173
Leonard 174
Deepak 182

13. Does the Brain Dictate Behavior? 189
Deepak 190
Leonard 198

14. Is the Brain Like a Computer? 205
Leonard 206
Deepak 216

15. Is the Universe Thinking Through Us? 227
Deepak 228
Leonard 234

Part Five: GOD

16. Is God an Illusion? 245
Deepak 246
Leonard 252

17. What Is the Future of Belief? 259
Deepak 260
Leonard 268

18. Is There a Fundamental Reality? 277
Leonard 278
Deepak 286

Epilogue 293
Leonard 294
Deepak 300

Acknowledgments 305
Index 307

Foreword

Nothing is more mysterious than another person's worldview. Each of us has one. We believe that our worldview expresses reality. The Native Americans of the Southwest traveled hundreds of miles to hunt buffalo but never ate fish from their local streams. In their worldview, it was real that fish were the spirits of departed ancestors. In the Old Testament it was real that animal sacrifices appeased God's wrath; to the everyday Roman it was real that the future could be foretold in the entrails of a chicken. To the ancient Greeks it was real that a moral individual could keep slaves and that there existed many gods, of love and beauty, war, the underworld, the hunt, the harvest, the sea.

What happens, then, when two worldviews clash? In 399 BCE three Athenian citizens accused Socrates of refusing to recognize the traditional gods and introducing new divinities instead (he was also accused of corrupting their youth). The penalty for this clash of worldviews, or gods, was death. During his trial Socrates refused to back down or to flee from a certain verdict of guilty. According to Plato, he said, "So long as I draw breath and have my faculties, I shall never stop practicing philosophy." Unfortunately, in many parts of the world today, a clash of worldviews is still met with violence and death.

This book is about a clash of worldviews, but no blows were exchanged. The book came about when two strangers met at a televised debate on "the future of God." The setting was an auditorium at the California Institute of Technology, and the audience was composed of many scientists and students, but also of laypeople, including Deepak's fans from the surrounding community. Each of them brought his or her own personal beliefs—no doubt some of them were religious—but they also brought their own worldview, which runs much deeper than belief.

In the Caltech debate Deepak served as the defender of a worldview broadly known as spiritual. Since the ideas of physics became an issue, during the question-and-answer period Deepak asked, "Is there a physicist in the house?" Neither Leonard nor anyone else answered. But after the debate, the moderator, who recognized Leonard as a physicist, pulled him out of the audience to ask Deepak a question. Leonard instead offered to teach him about quantum physics. Deepak accepted—to a mixture of laughter and applause—and as we started to communicate, we found ourselves strongly disagreeing about our worldviews. Realizing the depth of our clash, we decided to have it out in this book.

Science has set humanity on a path to unravel the secrets of nature, harness natural forces, and develop new technologies, using reason and observation instead of emotional bias as a tool for uncovering the truth of things. Spirituality looks toward an invisible, transcendent realm discovered within the self. Science explores the world as it is offered to the five senses and the brain, while spirituality considers the universe to be purposeful and imbued with meaning. In Deepak's view, the great challenge for spirituality is to offer something that science cannot provide—in particular, answers that lie in the realm of consciousness.

Which worldview is right? Does science describe the universe, or do ancient teachings like meditation unravel mysteries that are beyond the worldview of science? To find out, this book explores the clash of worldviews on three levels: the cosmos, or physical universe; life; and the human brain. Finally, we also explore the ultimate mystery, God. In "Cosmos" we argue about where the universe came from, its nature, and where it is going. In "Life" we debate evolution, genetics, and the origin of life. "Mind and Brain" addresses neuroscience and raises all the issues of mind and body. And "God" refers not only to a presiding deity but also to the broader concept of a divine presence in our universe.

This book covers eighteen topics in total, with essays from both authors. Each of us told his side of the story, one topic at a time, but whoever came second on any given topic did so with the other's text in hand, feeling free to present a rebuttal. Since rebuttals tend to persuade audiences, we tried to be as fair as possible about who got that advantage.

Each of us believes deeply in the worldview he represents. We have written fiercely but respectfully to define the truth as we see it. No one can ignore the question of how to perceive the world. The best we can do—writers and readers alike—is to leap into the fray. What else could be more important?

Deepak Chopra
Leonard Mlodinow

PART ONE

THE WAR

1

Perspectives

The Spiritual Perspective

DEEPAK

Who looks outside, dreams; who looks within, awakens.
—Carl Jung

If it is going to win the struggle for the future, spirituality must first overcome a major disadvantage. In the popular imagination, science long ago discredited religion. Facts replaced faith. Superstition was gradually vanquished. That's why Darwin's explanation of man's descent from lower primates prevails over Genesis and why we look to the Big Bang as the source of the cosmos rather than to a creation myth populated by one or more gods.

So it's important to begin by saying that religion isn't the same as spirituality—far from it. Even God isn't the same as spirituality. Organized religion may have discredited itself, but spirituality has suffered no such defeat. Thousands of years ago, in cultures across the globe, inspired spiritual teachers such as the Buddha, Jesus, and Lao-tzu proposed profound views of life. They taught that a transcendent domain resides beyond the everyday world of pain and struggle. Although the eye beholds rocks, mountains, trees, and sky, this is only a veil drawn over a vast, mysterious, unseen reality. Beyond the reach of the five senses lies an invisible realm of infinite possibility, and the key to unfolding its potential is consciousness. Go within, the sages and seers declared, and you will find the true source of everything: your own awareness.

It was this tremendous promise that religion failed to deliver on. The reasons don't concern us here, because this is a book about the future. It's enough to say that if the kingdom of God is within, as Christ declared, if nirvana means freedom from all suffering, as the Buddha taught, and if knowledge of the cosmos is locked inside the human mind, as the ancient *rishis,* or sages, of India proposed, we cannot look around today and say that those teachings bore fruit. Increasingly few people worship in the old ways around the world, and even as their elders lament this decline, those who have walked away from religion no longer even need an excuse. Science long ago showed us a brave new world that requires no faith in an invisible realm.

The real issue is knowledge and how you attain it. Jesus and the Buddha had no doubt that they were describing reality from a position of true knowledge. After more than two thousand years, we think we know better.

Science celebrates its triumphs, which are many, and excuses its catastrophes, which are also numerous—and growing. The atomic bomb delivered us into an age of mass destruction that brings night terrors just to contemplate. The environment has been disastrously disrupted by emissions spewing from the machines that technology gives us to make life better. Yet supporters of science shrug off these threats as either side effects or failures of social policy. Morality, we are told, isn't the responsibility of science. But if you look deeper, science has run into the same problem as religion. Religion lost sight of humility before God, and science lost its sense of awe, increasingly seeing Nature as a force to be opposed and conquered, its secrets stripped bare for the benefit of humankind. Now we are paying the price. When asked if *Homo sapiens* is in danger of extinction, some scientists offer hope that within a few hundred years space travel will be advanced enough to let us abandon the planetary nest we are fouling. Off we go to spoil other worlds!

We all know what's at stake: the foreseeable future looms grimly over us. The standard solution for our present woes is all too familiar. Science will rescue us with new technology—for restoring the environment, replacing fossil fuels, curing AIDS and cancer, and ending the threat of famine. Name your malady and there's someone to tell you that a scientific solution is just around the corner. But isn't science promising to rescue us from itself? And why is that a promise we should trust? The worldview that triumphed over religion, and that looks upon life as essentially materialistic, has set us on a path that leads to a dead end. Literally.

Even if we miraculously eliminated disastrous pollution and waste, coming generations will still have no model for the good life except the one that has failed us: endless consumption, exploitation of natural resources, and the diabolical creativity of warfare. As a young Chinese student bitterly commented about the West, "You ate the whole banquet. Now you give us coffee and dessert, but tell us to pay for the entire meal."

Religion cannot resolve this dilemma; it has had its chances already. But spirituality can. We need to go back to the source of religion. That source isn't God. It's consciousness. The great teachers who lived millennia ago offered something more radical than belief in a higher power. They offered a way of viewing reality that begins not with outside facts and a limited physical existence, but with inner wisdom and access to unbounded awareness. The irony is that Jesus, the Buddha, and the other enlightened sages were scientists, too. They had a way of uncovering knowledge that runs exactly parallel to modern science. First came a hypothesis, an idea that needed testing. Next came experimentation to see if the hypothesis was true. Finally came peer review, offering the new findings to other researchers and asking them to reproduce the same breakthrough.

The spiritual hypothesis that was put forward thousands of years ago has three parts:

1. There is an unseen reality that is the source of all visible things.
2. This unseen reality is knowable through our own awareness.
3. Intelligence, creativity, and organizing power are embedded in the cosmos.

This trio of ideas is like the Platonic values in Greek philosophy, which tell us that love, truth, order, and reason shape human existence from a higher reality. The difference is that even more ancient philosophies, with roots going back five thousand years, tell us that higher reality is with us right here and now.

In the following pages, as Leonard and I debate the great questions of human existence, my role is to offer spiritual answers—not as a priest or a practitioner of any particular faith, but as a researcher in consciousness. This runs the risk, I know, of alienating devout believers, the many millions of people in every faith for whom God is very personal. But the world's wisdom traditions did not exclude a personal God (to be candid, I was not taught as a child to worship one, but my mother did, praying at a temple to Rama every day of her life). At the same time, wisdom traditions all included an impersonal God who permeates every atom of the universe and every fiber of our being. This distinction bothers those believers who want to cling to the one and only true faith, whatever it may be for them. But an impersonal God doesn't need to be a threat.

Think of someone you love. Now think of love itself. The person you love puts a face on love, yet surely you know that love existed before this person was born and will survive after they pass away. In that simple example lies the difference between the personal and the impersonal God. As a believer you can put a face on God—that is a matter of your own private choice—but I hope you see that if God is everywhere, the divine qualities of love, mercy, compassion, justice, and all the other attributes ascribed to God extend infinitely throughout creation. Not surprisingly, this idea is a common thread in all major religions. Higher consciousness allowed the great sages, saints, and seers to attain a kind of knowledge that science feels threatened by but that is completely valid. Our common understanding of consciousness is too limited to do justice here.

If I asked you, "What are you conscious of right this minute?" you would probably start by describing the room you're in and the sights, sounds, and smells surrounding you. On reflection you'd become aware of your mood, the sensations in your body, perhaps a hidden worry or desire that lies deeper than superficial thoughts. But the inner journey can go much deeper, taking you to a reality that isn't about objects "out there" or feelings and thoughts "in here." Eventually those two worlds meld into one state of being that lies beyond the limits of space-time, in a realm of infinite possibilities.

Now we face a contradiction, however. How can two realities that are opposites (the way baking a loaf of bread is the opposite of dreaming about a loaf of bread) turn out to be the same? This improbable vision is succinctly described in the Isha Upanishad, an ancient Indian scripture. "*That* is complete, and *this* is also complete. *This* totality has been projected from *that* totality. When *this* wholeness merges in *that* wholeness, all that remains is wholeness." At first glance, this passage seems like a riddle, but it can be deciphered by realizing that *"that"* is the state of pure consciousness, while *"this"* is the visible universe. Both are complete in themselves, as we know from science, which has been satisfied for four centuries with exploring the visible universe. But in the spiritual worldview a hidden wholeness underlies all of creation, and ultimately it is this invisible wholeness that matters most.

Spirituality has been around for many thousand years, and its researchers were brilliant—the very Einsteins of consciousness. Anyone can reproduce and verify their results, as with the principles of science. More important, the future that spirituality promises—one of wisdom, freedom, and fulfillment—hasn't vanished as the age of faith declined. Reality is reality. There is only one, and it's permanent. This means that at some point the inner and outer worlds must meet; we won't have to choose between them. That in itself will be a revolutionary discovery, since the dispute between science and religion has persuaded almost everyone that either you face reality and deal with the tough questions of everyday life (science), or you passively retreat and contemplate a realm beyond everyday life (religion).

This either/or choice was forced on us when religion failed to deliver on its promises. But spirituality, the deeper source of religion, hasn't failed and is ready to meet science face-to-face, offering answers consistent with the most advanced scientific theories. Human consciousness created science, which ironically is now moving to exclude consciousness, its very creator! Surely this would leave us with worse than an orphaned and shrunken science—we'd inhabit an impoverished world.

It has already arrived. We live in a time of rude atheism, whose proponents deride religion as superstition, illusion, and a hoax. But their real target isn't religion; it's the inner journey. I am less concerned with attacks on God than I am with a far more insidious danger: the superstition of materialism. To scientific atheists, reality must be external; otherwise their whole approach falls apart. If the physical world is all that exists, science is right to mine it for data.

But here the superstition of materialism breaks down. Our five senses encourage us to accept that there are objects "out there," forests and rivers, atoms and quarks. However, at the frontiers of physics, where Nature becomes very small, matter breaks down and then vanishes. Here, the act of measuring changes what we see; every observer turns out to be woven into what he observes. This is the universe already known to spirituality, where passive observation gives way to active participation, and we discover that we are part of the fabric of creation. The result is enormous power and freedom.

Science has never achieved pure objectivity, and it never will. To deny the worth of subjective experience is to dismiss most of what makes life worth living: love, trust, faith, beauty, awe, wonder, compassion, truth, the arts, morality, and the mind itself. The field of neuroscience has largely accepted that the mind doesn't exist but is merely a by-product of the brain. The brain (a "computer made of meat," as Marvin Minsky, an expert in artificial intelligence, dubbed it) is our master, chemically deciding how we feel, genetically determining how we grow, live, and die. This picture isn't acceptable to me, because in dismissing the mind we eliminate our portal to knowledge and insight.

As Leonard and I debate the big mysteries, the great sages and seers remind us that there is only one question: What is reality? Is it the result of natural laws rigorously operating through cause and effect, or is it something else? There is good reason for our worldviews to be at war. Either reality is bounded by the visible universe, or it isn't. Either the cosmos was created from an empty, meaningless void, or it wasn't. Until you understand the nature of reality, you are like one of the fabled six blind men trying to describe an elephant by holding on to just one of its parts. The one who has hold of the leg says, "An elephant is much like a tree." The one who has hold of the trunk says, "An elephant is much like a snake." And so on.

The childhood fable about the blind men and the elephant is actually an allegory from ancient India. The six blind men are the five senses plus the rational mind. The elephant is Brahman, the totality of all that exists. On the surface the fable is pessimistic: if all you possess is your five senses and your rational mind, you'll never see the elephant. But there is a hidden message so obvious that many people miss it. The elephant exists. It was there before us, patiently waiting to be known. It is the deeper truth of unified reality.

Just because religion didn't succeed doesn't mean that a new spirituality, based on consciousness, won't. We need to see the truth, and in the process we will awaken the profound powers that were promised to us thousands of years ago. Time awaits. The future depends on the choice we make today.

The Scientific Perspective

LEONARD

The further the spiritual evolution of mankind advances, the more certain it seems to me that the path to genuine religiosity does not lie through the fear of life, and the fear of death, and blind faith, but through the striving after rational knowledge.
—ALBERT EINSTEIN

Children come into the world believing it all revolves around them, and so did humanity. People have always been anxious to understand their universe, but for most of human history we hadn't yet developed the means. Since we are proactive and imaginative animals, we didn't let the lack of tools stop us. We simply applied our imagination to form compelling pictures. These pictures were not based on reality, but were created to serve our needs. We would all like to be immortal. We'd like to believe that good triumphs over evil, that a greater power watches over us, that we are part of something bigger, that we have been put here for a reason. We'd like to believe that our lives have an intrinsic meaning. Ancient concepts of the universe comforted us by affirming these desires. Where did the universe come from? Where did life come from? Where did people come from? The legends and theologies of the past assured us that we were created by God, and that our Earth was the center of everything.

Today science can answer many of the most fundamental questions of existence. Science's answers spring from observation and experiment rather than from human bias or desire. Science offers answers in harmony with nature as it is, rather than nature as we'd like it to be.

The universe is an awe-inspiring place, especially for those who know something about it. The more we learn, the more astonishing it seems. Newton said that if he saw further it was because he stood on the shoulders of giants. Today we can all stand on the shoulders of scientists and see deep and amazing truths about the universe and our place in it. We can understand how we and our Earth are natural phenomena that arise from the laws of physics. Our ancestors viewed the night sky with a sense of wonder, but to see stars that explode in seconds and shine with more light than entire galaxies brings a new dimension to the awe. In our day a scientist can turn her telescope to observe an Earthlike planet trillions of miles away, or study a spectacular internal universe in which a million million atoms conspire to create a tiny freckle. We know now that our Earth is one world among many and that our species arose from other species (whose members we may not wish to invite into our living rooms but who are our ancestors nonetheless). Science has revealed a universe that is vast, ancient, violent, strange, and beautiful, a universe of almost infinite variety and possibility, one in which time can end in a black hole, and conscious beings can evolve from a soup of minerals. In such a universe it can seem that people are insignificant, but what is significant and profound is that we, ensembles of almost uncountable numbers of unthinking atoms, can become aware, and understand our origins and the nature of the cosmos in which we live.

Deepak feels that scientific explanations are sterile and reductive, diminishing humankind to a mere collection of atoms, no different in kind from any other object in the universe. But scientific knowledge does not diminish our humanity any more than the knowledge that our country is one among many diminishes our appreciation of our native culture. In fact, the opposite is closer to the truth. Emotion, intuition, adherence to authority—traits that drive the belief in religious and mystical explanation—are traits that can be found in other primates, and even in lower animals. But orangutans cannot reason about the angles in triangles, and macaque monkeys do not look to the heavens and wonder why the planets follow elliptical paths. It is only humans who can engage in the wondrous processes of reason and thought called science, only humans who can understand themselves and how their planet got here, and only humans who could discover the atoms that form us.

The triumph of humanity is our capacity to understand. It is our comprehension of the cosmos, our insight into where we came from, our vision of the place we occupy in the universe, that sets us apart. A by-product of this scientific understanding is the power to harness nature for our benefit, or, it is true, to employ it to our detriment. The particular ethical and moral choices people make depend on human nature, and human culture. People dropped boulders on their enemies long before they understood the law of gravity. And they spewed filth into the skies long before they understood the thermodynamics of burning coal.

Promoting good and avoiding evil is the charge of organized religion and spirituality. It is those enterprises—not science—that have often failed to deliver on their promise. Eastern religions did not prevent a history of brutal warfare in Asia, nor did Western religions pacify Europe. In fact, more people have been slaughtered in the name of religion than by all the atomic weapons made possible by modern physics. From the Crusades to the Holocaust, in addition to being a tool of goodness and love, religion has been employed as a tool of hatred. Deepak's universalist and peaceful approach to spirituality is therefore a welcome alternative. But Deepak's metaphysics goes beyond spiritual guidance to offer views on the nature of the universe. Deepak's belief that the universe is purposeful and imbued with love may be attractive, but is it correct?

Deepak criticizes science for its vision of life as "essentially materialistic." By materialistic, Deepak does not mean to suggest that scientists are focused only on things and the desire to possess them, but that scientists deal only with phenomena we can see, hear, smell, detect with instruments, or measure with numbers. He contrasts the visible, or detectable, universe studied by science with an implicitly superior but invisible "realm of infinite possibility" that lies beyond our senses, a "transcendent domain" that is the source of all visible things. Deepak argues passionately that only by accepting this realm can science grow beyond its limits and help save the world. But arguing that such a realm can expand the limits of science, that it can help humanity, or that ancient sages taught about it doesn't make it true. If you think you are eating a cheeseburger, and I tell you that in some other unseen realm it is really a filet mignon, you'd want to know how I know this, and what evidence supports my idea. Only those answers can enable a belief to transcend wish fulfillment, so if Deepak is to be convincing, those questions are the challenges he must address.

The real issue, as Deepak says, is knowledge and how you attain it. Deepak criticizes science for denying "the worth of subjective experience." But science wouldn't have gotten very far if one scientist described a helium atom as "pretty heavy" while another noted that "it feels light to me." Scientists employ precise objective measurements and precise objective concepts for good reason, and the fact that they seek to ensure that their measurements and concepts are not influenced by "love, trust, faith, beauty, awe, wonder, compassion," etc., does not mean that they dismiss the value of those qualities in other areas of life.

Scientists are often guided by their intuition and subjective feelings, but they recognize the need for another step: verification. Science proceeds in a loop of observation, theory, and experiment. The loop is repeated until the theory and the empirical evidence are in harmony. But this method would fail if concepts were not precisely defined and experiments were not rigorously controlled. These elements of the scientific method are crucial, and it is they that determine the difference between good science and bad science, or between science and pseudoscience. Deepak said Jesus was a scientist. Was he? He probably did not gather a sample of the population and, after being insulted, turn the other cheek to half of them, and lay out the other half with a solid right hook, then gather statistics on the efficacy of the different approaches. It might seem silly that I object when Deepak calls Jesus a scientist, but it introduces a theme—the use of terminology—that will become important in more substantive contexts later in this book: one must be careful when discussing scientific issues not to use terms loosely. It is easy to use words imprecisely in an argument, but it is also dangerous, because the substance of the argument often relies on the nuances of those words.

I do not suggest that science is perfect. Deepak says that science has never achieved pure objectivity, and he is right. For one, the concepts employed in science are concepts conceived by the human brain. Aliens with different brain structures, thought processes, and sense organs might view matter in completely different, but equally valid, ways. And if there is a certain kind of subjectivity to our concepts and our theories, there is also subjectivity in our experiments. In fact, experiments that have been done on experimenters show that there is a tendency for scientists to see what they want to see, and to be convinced by data they wish to find convincing. Yes, scientists, and science, are fallible. Yet all these are reasons not to doubt the scientific method, but to follow it as scrupulously as possible.

History shows that the scientific method works. Being only human, some scientists may at first resist new and revolutionary ideas, but if a theory's predictions are confirmed by experiment, the new theory soon becomes mainstream. For example, in 1982, Robin Warren and Barry Marshall discovered the *Helicobacter pylori* bacteria, and hypothesized that it causes ulcers. Their work was not well received because at the time scientists firmly believed that stress and lifestyle were the major causes of peptic ulcer disease. Yet further experiments bore out their claims, and by 2005 it had been established that *Helicobacter pylori* causes more than 90 percent of duodenal ulcers and up to 80 percent of gastric ulcers, and Warren and Marshall were awarded the Nobel Prize. Science would also embrace Deepak, if his claims were true.

When theories that people are passionate about are brushed off by the science community, cries of closed-mindedness often emerge. But the history of science shows that the real reason for the rejection of theories is that they clash with observational evidence. In fact, some very weird ideas, arising sometimes from very obscure and unexpected quarters—ideas like relativity and quantum uncertainty—have quickly gained acceptance, despite challenging conventional thinking, for just one reason: they passed their experimental tests. Proponents of metaphysics and Deepak's spirituality are far less open to revising or expanding their worldviews to encompass new discoveries. Rather than welcoming new truths, they often cling to ancient ideas, explanations, and texts. If on occasion they turn to science in an attempt to justify their traditional ideas, whenever it appears that science does not support them they are quick to turn their backs on it. And when they do employ scientific concepts, they use them so loosely that the meanings are altered, with the result that the conclusions they come to are not valid.

One can't expect science to answer all the questions of the universe. There may well be secrets of nature that will remain forever beyond the outer limits of human intelligence. Other questions, such as those regarding human aspirations and the meaning of our lives, are best viewed from multiple perspectives, both scientific and spiritual. These approaches can coexist and respect each other. The trouble arises when religious and spiritual doctrine makes pronouncements about the physical universe that contradict what we actually observe to be true.

To Deepak, the key to everything is the understanding of consciousness. It is true that science has only begun to address that question. How do those unthinking atoms we are made of conspire to create love, pain, and joy? How does the brain create thought and conscious experience? The brain contains more than a hundred billion neurons, roughly the number of stars in a galaxy, but the stars hardly interact, while the average neuron is plugged into thousands of others. That makes the human brain far more complex and difficult to fathom than the universe of galaxies and stars, and is one reason we have made great leaps in our understanding of the cosmos, while knowledge of ourselves proceeds at a relative crawl. Is that a sign that our minds cannot be explained?

It is shortsighted to believe that because science today cannot explain consciousness, consciousness must lie beyond science's reach. But even if the origin of consciousness *is* too complex to be fully grasped by the human mind, that is not evidence that consciousness resides in a supernatural realm. In fact, though the question of how consciousness arises remains a puzzle, we have plenty of evidence that consciousness functions according to physical law. For example, in neuroscience experiments, thoughts, feelings, and sensations in subjects' minds—the desire to move an arm, the thought of a specific person like Jennifer Aniston or Mother Teresa, and the craving for a Snickers bar—have all been traced to specific areas and activities in the physical brain. Scientists have even uncovered what they call "concept cells," which fire whenever a subject recognizes a concept, such as a specific person, place, or object. These neurons are the cellular substrate of an idea. They will fire, say, each time a person recognizes Mother Teresa in a photo, no matter what her dress or pose. They will even fire if the subject merely sees her name spelled out in text.

Science can answer the seemingly intractable question of how the universe came into being, and there is reason to believe that science will eventually be able to explain the origins of consciousness, too. Science is an ever-advancing process, and the end is not in sight. If at some future date we *are* able to explain the mind in terms of the activity of a universe of neurons, if all our mental processes *do* prove to have their source in the flow of charged ions within nerve cells, that would not mean that science denies the worth of "love, trust, faith, beauty, awe, wonder, compassion, truth, the arts, morality, and the mind itself." To explain something is not, as I have said, to diminish or deny its worth. It is also important to recognize that even if we consider a scientific explanation of our thought processes (or anything else) aesthetically or spiritually unsatisfying or unpalatable, that does not make it false. Our explanations must be guided by truth; truth cannot be adjusted to conform to what we want to hear.

Unfortunately, the current absence of a fully developed scientific theory of consciousness invites just the type of imprecise reasoning that leads to conclusions that conflict with known physical laws. Philosophy and metaphysics cannot explain an MRI machine, a television, or even a toaster. Can they explain consciousness, or why the universe is as we find it? Maybe, but as Deepak offers his explanations of a universal consciousness, I plan to hold to an important principle of science, skepticism. Deepak tells me that in our discussion he is the underdog. The data show otherwise. According to random samples, only 45 percent of the American public believes in evolution, but 76 percent believes in miracles. No presidential candidate can be credible without proclaiming a belief in some higher power, but many have found it politically advantageous to deny the theory of evolution. Science is not the lord of modern life Deepak imagines, but its underappreciated servant.

The answers of science don't come easily. Nobel Prize–winning physicist Steven Weinberg has dedicated his life to the tireless study of the theory of elementary particles, such as the electron, the muon, and the quark. Yet he wrote that he has never found those particles very interesting. Why then has he devoted his life to understanding them? Because he believes that at this moment in the history of human thought, their study offers the most promising way to achieve insight into the fundamental laws that govern all of nature. Some of the ten thousand scientists who worked, many for over a decade, to build the Large Hadron Collider, the multibillion-dollar particle accelerator in Geneva, probably didn't think the long hours of calibrating delicate instruments and fine-tuning spectrometers was all that fascinating either (though many certainly did!). They did it for the same reason Weinberg studied muons. Humans are unlike other animals in the questions they ask about their environment. When dropped into new surroundings, a rat will explore for a while, form a mental map, get safe, then stop probing. But a person will ask, Why am I in this cage? How did I get here? Where's the nearest decent coffee? Humans study science because we have an urge to know how our lives fit into the greater scheme of the universe. That's one of the defining qualities of what makes us human. But the answers are only edifying if they are true. So to you, the reader, I would suggest that as you ponder Deepak's often very appealing worldview, you keep in mind the words of the iconic Caltech physicist Richard Feynman: the first principle is that you must not fool yourself—and you are the easiest to fool.

PART TWO

COSMOS

2

How Did the Universe Emerge?

LEONARD

Every civilization has had its creation stories. The Europeans came up with a doozy in the early twentieth century, and it has since been refined and elaborated upon by scholars from all over the world. It came to be called the Big Bang, but it has morphed into something called the standard model of cosmology. We consider it a theory, while we call the other explanations myths. What makes the Big Bang different from the Mayan proposition that we are all made from white and yellow corn? Is science's faith in its explanation justified? What are the limits of current knowledge?

The idea of the Big Bang arose from Einstein's theory of general relativity, which he completed in 1915, after over a decade of work. General relativity is a set of equations that describe the way gravity, space and time, energy and matter, all interact. With his theory Einstein was asking people to toss out the intuitively satisfying and very successful theory of Isaac Newton, and in its place to accept some very weird ideas that seem to contradict what we experience in everyday life. Metaphysics is a court of opening and closing arguments, with no requirement that evidence be presented in between. In science it is only the evidence that matters. So when Einstein said there is a hidden reality underlying and quite different from the world we perceive with our senses, no scientist would have listened unless he produced a series of smoking guns. He did.

Though one can apply general relativity to the universe as a whole, the applications that provide the easiest tests of its validity are the ones that successfully explain simple systems such as a planet orbiting our sun, or a ray of starlight flying past it. It was these applications that provided the first physical evidence that Einstein was onto something. In the case of the planet, Einstein's theory explained a previously observed irregularity in the orbit of Mercury, which deviated from the prediction of Newton's laws. It was a small irregularity, so most scientists before Einstein had simply scratched their heads over it, and expected that eventually a mundane explanation would be found. Einstein showed that the explanation was anything but mundane. Because that irregularity was already known, an even more impressive test of the theory was his novel (and at the time astonishing) prediction that, given the effects of relativity, gravity would bend light rays, and hence that our view of distant stars would be altered when their light passed near our sun. In order to observe that effect, and not have the starlight in question swamped by that of the sun, one had to look at it during a total solar eclipse. This experiment was performed, and Einstein's theory was found to correctly predict not just that the light would be bent, but also the amount of the bending.

Einstein's triumph—and the equally revolutionary triumph of quantum theory—did not mean that everything about Newton's view of the world had suddenly been invalidated. It is not as if civilization woke up one morning and realized it had built all its buildings and bridges wrong, that Edison's lightbulb is really a quantum laser, or that if you drive faster than the speed limit you'll never need wrinkle cream. Newton's theory had been tested and retested, and, except for the problem of the orbit of Mercury, never been found lacking, and Einstein's theory didn't challenge the fact that Newton's theory provides an excellent description of the events we experience in our everyday lives. In fact, when applied to such situations, Einstein's theory yields predictions so close to those of Newton's that only very sophisticated instruments can detect the difference. But under certain conditions, relevant for astrophysics and in certain laboratory experiments, Newtonian predictions do differ significantly from those of Einstein's theory. So when scientists say that Newton's theory is "wrong," we mean it is only approximately correct. Still, Einstein's theory is a more fundamentally true description of nature, which reveals the character of space and time on a much deeper level than what Newton had envisioned.

The experimental support for his theories made Einstein an international celebrity, but the most astounding implications of his ideas were yet to come. In the 1920s a Belgian priest and astronomer named Georges Lemaître applied Einstein's equations to the universe as a whole. He discovered something that at the time might have seemed both obvious and shocking. First the obvious part. Since gravity is an attractive force, when you toss an apple into the air, the pull of gravity will cause it to fall back toward the Earth. That is, the apple first moves away from the earth, then back down toward it, but does not hover in place (except for that single instant at the top of its trajectory). The shocking part came when Lemaître showed that, similarly, due to the mutual attraction of the matter and energy within it, the universe can expand, slow down, and possibly contract, but cannot remain at a fixed size, as everyone at the time—including Einstein—believed. If the universe is expanding, that means that if you trace the history of the universe backward in time, you'll find the universe getting ever smaller. And so Lemaître speculated further that the universe began as a single point. That theory is now called the Big Bang theory.

The Big Bang theory was intimately connected to Einstein's general relativity, but if it had made no testable predictions it would have been little better than saying the universe was made from corn. A critical element of the theory was confirmed shortly after Lemaître's work, when Edwin Hubble discovered that the universe *is* expanding. But a more specific implication of Lemaître's scenario is that, as the primordial fireball cooled to a billion degrees in the first few minutes after the Big Bang, various light elements should have been created in certain definite proportions. In particular, about 25 percent of the matter in the universe should be in the form of helium—and this is precisely what we find. Another implication is that the universe should have cooled a great deal more since then. According to the theory, space today should be permeated with residual radiation at a temperature of, on average, about 2.7 degrees centigrade above absolute zero. Again, this agrees with what we measure.

By the 1970s the Big Bang model had proved very successful at explaining most of the history of our universe. But there remained some apparent anomalies. For example, consider a frying pan that is at a uniform temperature except for one spot that is hotter than the rest. After a short time, the hot spot will be a bit cooler, while the nearby region of the pan will be slightly warmer. With more time, the hot spot will cool further, transferring its heat to ever larger areas of the pan. Eventually the entire pan will end up at a uniform temperature. But this transition to uniformity takes time. The universe is like the pan after a very long time—its temperature is almost uniform. The problem was that we happened to know that not enough time had passed to have allowed that to occur. So why is it so close to 2.7 degrees in every direction? Why not a hot spot here and a cold spot there? Physicists called this the horizon problem.

The so-called flatness problem was another puzzle. General relativity dictates that the amount of matter and energy in the universe determines the curvature of space. What does that mean? Curvature of our three-dimensional space can be difficult to visualize, but the idea is similar in two dimensions, so let's consider that. A flat plane is a two-dimensional surface with no curvature. The surface of a sphere, on the other hand, curves in on itself, and is an example of a surface with what is called positive curvature. In contrast, a saddle is curved outward, so it is said to have negative curvature. The equations of general relativity tell us that if there is more than a certain critical amount of matter and energy per unit volume in the universe, space will curl up into a spherelike shape, and eventually collapse upon itself. If there is less than this critical density, space will curve outward like a saddle. Only if the average concentration of matter and energy is exactly at the critical value will space be flat. The critical density varies with the age of the universe. Long ago it was very high, but today it is the equivalent of about 6 hydrogen atoms per cubic meter of space.

We can measure the large-scale curvature of space directly, and space appears to be flat, at least to the precision to which we can measure. The problem is that the equations of general relativity show that if the density of the universe ever deviated from the critical value, that deviation would quickly get enormously amplified. That means that if, in the early universe, the density of matter had been even slightly less than the critical density, the universe would today be saddle-shaped and vastly more dilute than we find it. Or if its density had been just a bit higher than the critical value, the universe would long ago have collapsed in on itself like a balloon with the air sucked out. Due to this amplification effect, in order for the Big Bang model to account for the degree of flatness that we observed, when the universe was one second old, the concentration of matter and energy had to be tuned to the critical value within an accuracy of one part in a thousand trillion.

One might ask, "So what? Couldn't the universe simply have been made that way?" It could have, but this illustrates an important point in science. The key aspects of a theory should follow from some principle, and not be contrived to make the theory work. To a scientist, a theory stating that the universe depends upon being set up long ago in a very precise way is not a very satisfying theory. Scientists want to comprehend the underlying reason, the natural laws that explain the special circumstance.

The horizon problem, the flatness problem, and some other difficulties with the Big Bang theory were all resolved in the late 1970s when physicists discovered a new chapter in the evolution of the universe, a chapter called inflation. Inflation was discovered by Alan Guth, a young particle theorist who, by his own admission, hadn't really accomplished very much up until then. Guth changed that when he realized that certain conditions that physicists believe were present when the universe was a fraction of a second old would have caused the cosmos to go crazy, doubling in size in less than every billionth of a trillionth of a trillionth of a second. Assuming that doubling continued for "only" a hundred cycles, a parcel of universe the width of a penny would have blown up to more than ten million times the diameter of the Milky Way.

How does inflation benefit a troubled cosmologist? Imagine running a film of the universe backward from today. As we move backward through inflation, the observable universe is crunched into an extremely tiny region. Inflation therefore means that regions of the universe that are now widely separated were close enough together in preinflation times that their temperature differences could have been smoothed out before the expansion. That solves the horizon problem. Inflation also solves the flatness problem. To understand why, imagine what would happen to a tiny balloon that suddenly inflated to the diameter of, say, the sun. Though it would have been easy to measure the balloon's curvature before its great blowup, once it is the size of the sun, to anyone on its surface, the balloon would appear much flatter. In an analogous way, inflation flattened our universe.

Guth's theory could not have been envisioned by Einstein, Lemaître, or anyone else working with general relativity alone. It depended on ideas taken from that other revolution of the twentieth century, quantum theory. Quantum theory is not really a theory, but a set of principles that define a type of theory. Theories developed according to those quantum principles are called quantum theories. General relativity is not a quantum theory and we don't yet know exactly how to make it one, but there are ways of extracting limited predictions that draw upon the principles of both theories. In his work Guth relied on many quantum ideas developed between the 1930s and the 1970s.

One of the basic tenets of any modern quantum theory is that for every particle there is a field, something like the force fields you see in science fiction. According to quantum theory, those fields cannot remain constant in magnitude, but are subject to continual quantum fluctuations on a microscopic scale. As inflation began to occur and the old wrinkles in space were flattened out, new microscopic quantum wrinkles arose to replace them. As inflation progressed, it stretched those wrinkles to macroscopic size, resulting in a specific pattern of variation in the matter/energy density of the postinflationary universe. And since gravity is attractive, the areas that emerged from inflation denser than their surroundings attracted ever more matter, creating the seeds of galaxies. In that way the stretched-out quantum fluctuations led to the structure we see in the universe today—the galactic clusters, galaxies, and stars. Without the quantum fluctuations, the universe would be a uniform and featureless soup.

The pattern of density variation created by inflation can still be detected today. Earlier I said the fact that the temperature of the universe was almost the same everywhere was a mystery that inflation explains. But inflation goes a step further—it predicts that although the temperature is *nearly* constant in any direction you look, it will vary slightly and in a particular pattern. That is a very precise prediction, and a high bar of evidence to clear, but temperature variations of exactly the kind predicted by inflation have now been observed, variations that occur within a range of less than a hundred thousandth of a degree centigrade.

That, in brief, is the scientific picture of how the universe got here, and some of the evidence for that scenario. The real bang in the Big Bang was not the beginning of the universe, but the period of inflation, an expansion many times more drastic than that predicted by the original Big Bang scenario, and one that happened an instant after the universe began.

What happened before inflation? For now, scientific answers to that question are far more speculative, and far less certain, than the picture I've described above. Better answers await progress in creating a quantum version of general relativity (string theory, if shown to be true, would accomplish that). Many physicists argue that the new theory, once we have it, will show that, at some point before inflation, time as we know it did not exist. But the most striking speculation about what a quantum theory that includes general relativity might tell us comes from a quantum principle called vacuum fluctuations.

I mentioned above that galaxies are products of the microscopic fluctuations of quantum fields. Vacuum fluctuations refer to the quantum prediction that even "nothingness"—which in quantum theory is given a precise mathematical definition—exhibits fluctuations, and is therefore in a sense unstable. That is, even if you start with a region of space in which there is neither energy nor matter, it will not remain that way. Nothingness is instead like a boiling cauldron in which particles are always bubbling in and out of existence. That is a strange concept taken in the context of everyday experience, but to those who spend their days studying the behavior of elementary particles, it is a familiar effect. Vacuum fluctuations are one of the best-confirmed results in all of science, and have been measured to an accuracy of ten decimal places. They must be accounted for in all calculations and experiments in modern particle physics. In fact, most of your mass comes from the protons in the atoms you are made of, and most of the mass of a proton comes, not from the masses of the quarks that make up the proton, but from the energy of the "empty" space between those quarks, the turbulent brew of particles arising from nothingness, and then quickly disappearing back into it. So next time you think about how much you weigh, remember that most of your weight is due to the weight of empty space.

Many physicists believe that vacuum fluctuations point to an astounding prediction: the universe could have arisen spontaneously from nothing. Did it? We don't yet know for sure because we don't yet understand exactly how general relativity and quantum theory can be combined. Even once we think we have figured it out, specific predictions pertaining to observable phenomena will have to be made, and those predictions tested. Physicists will do that, because that, ultimately, is the work of science. Unlike philosophical, metaphysical, and mystical speculations, which are not bound by the constraint of evidence, a scientific theory of the origin of the universe must pass observational tests. The resulting picture might not satisfy those looking for a divine source for our beginnings, but it will be the answer of science.

DEEPAK

The first and greatest mystery is how the universe came to be. For spirituality, the issue seems like a lost cause before discussion even begins. Modern physics has taken over the genesis question, and its answer—the Big Bang and all that followed for the next 13.7 billion years—has succeeded in wiping out the credibility of the Bible, the Koran, the Vedas, and every other indigenous version of creation. Yet today, just at the moment when science seems poised to strike the final blow, it has gotten stuck. Quantum physics has been forced to stop at the edge of the void that preceded creation, with no way forward until that void can be bridged by an explanation. Leonard's position, shared by physics in general, is that the full explanation will be found through mathematics. My position, shared by students of consciousness in general, is that the very meaning of existence is at stake. In modern times we have assigned cosmology to specialists the way we assign genes to geneticists. But you can't pin a sign on creation that says "Keep out; you don't know enough math." We all have a stake in genesis, and that's fortunate, because a new creation story is trying to be born in our time, and all previous versions are up for radical revision.

The void is the starting point for any creation story, whether scientific or spiritual. The book of Genesis tells us that "the earth was without form and void, and darkness was over the face of the deep." Assigning God a home in the void doesn't satisfy the scientific mind, however, and spirituality must overcome some strong skeptical objections, which include the following:

- There is no scientific proof that God, or any creator, exists.
- The universe cannot be proven to have a purpose.
- The preuniverse may be unimaginable. Insofar as our experiences happen in time and space, is it futile to try to explain reality before space and time appeared?
- Randomness seems to be the long-term winner in the universe as stars die and energy approaches absolute zero.

These seem like crushing objections, and Leonard exemplifies the stubborn resistance of science to other ways of regarding the cosmos. Nonscientific explanations he regards with suspicion or worse—as primitive superstition ("white and yellow corn"), or self-delusion. For him, all processes in the cosmos, visible or invisible, can be explained through materialism. But it's fascinating to see just how spirituality has resurged in the debate, and why, in my view, it will gain the upper hand. All of science's objections can be met, and in the process we can lay the groundwork for a new creation story.

Stephen Hawking is regarded by popular culture as the last wise man, who, like Einstein, carries the full weight of science when he speaks. Hawking made worldwide news in 2010 by declaring that "it is not necessary to invoke God . . . to set the universe going." The world of devout believers had one more reason to consider science the enemy of faith. Einstein personally felt awe and wonder before the mystery that lies at the far horizon of the cosmos. But since then the universe of theoretical physics has become random, complex, paradoxical, and barren of divine presence.

Hawking and others say that quantum principles make it possible for the universe to arise from nothingness. But to keep this from being the void that begins Genesis, physics ties itself into a knot. "Nothing" gave rise to the human desire for meaning, so how can it be meaningless? The universe operates randomly, but this randomness created the human brain, which does all kinds of nonrandom things (such as writing Shakespeare and saying "I love you"), so how did the purposeless give birth to the purposeful?

The still unproven nature of "nothing" is the opening wedge for spirituality, which, contrary to what Leonard says, doesn't need to revert to prescientific myths. Instead, it presents insights about what lies beyond space and time. The new creation story will be based on the following:

1. Wholeness: The universe, including the void that precedes creation, is one system. The ground of existence is not inert emptiness, but a dynamic field connecting all creation in a single totality. Smaller processes in the quantum field hang together even when they are light-years apart. We see all kinds of things happening around us that cannot be totally disconnected: How is a firefly on a sultry summer night connected with emperor penguins marching hundreds of miles across the Antarctic ice, or with a tropical storm in the East Indies? The deeper truth is that wholeness must include all of them.

Our five senses are caught up in diversity, and part of diversity's job is to look disconnected; that's what fascinates us about life's endless variety. Wholeness, on the other hand, is invisible. It can be fully known only with the probing mind in its deepest explorations—that is the spiritual perspective. The only external way to glimpse wholeness is through mathematics. As Einstein observed, he thought up the concept of relativity in mathematical terms and then was astonished that Nature agreed with him. But an inner experience of wholeness, which is what the Buddha and other sages report, is just as valid a form of knowledge, and in the end is more satisfying, as I hope to establish.

2. Orderliness: The natural laws that govern the universe are orderly because they can be mathematically explained. Events that look random, from the scattering of light to the bombardment of atoms, from wind to volcanic eruptions, distract us from that deeper truth: Randomness is just a way to get from one stage of orderliness to another. To put it another way, randomness is the universe's way of breaking eggs to make cosmic omelets. As higher orders of organization emerge, they go through messy transitions that seem to behave randomly—the way vegetables in a compost heap go through decay in order to become fertile soil—but randomness is not the end stage; it is only the intermediate step for a new, more complex level of organization. It is only a step from orderliness to meaning, which implies that the universe actually means something.

3. Evolution: The first cousin of randomness is entropy, the law of which states that heat is constantly being dispersed throughout the universe. Entropy is how the cosmos winds down, heading for absolute zero, to the so-called heat death that awaits all things. But another force exists that creates the opposite—warm spots in creation where heat collects, leading to DNA and life on Earth. This opposing force is evolution, the tendency that makes everything grow. Spirituality holds that evolution is dominant in Nature. Growth, once it begins, never ends.

4. Creativity: Evolution doesn't simply scramble old ingredients into new forms; nor does it just turn small clumps of matter into bigger clumps. Instead, evolution makes leaps of creativity. These happen in quantum form—that is, there is a sudden emergence of a property that never existed before. Water emerges from two invisible gases, hydrogen and oxygen. Nothing about those two gases would predict what water is like. Quantum leaps dominate in creation everywhere we look, but especially in the startling, beautiful novelty of life-forms on Earth. The cosmos is ruled by creativity.

5. Consciousness: To be creative you have to be conscious. Spirituality holds that consciousness is basic to creation. It has always existed, and the visible universe unfolds as a display of what consciousness wants to explore. Wholeness couldn't unfold simply by following mechanical laws such as gravity. Looking around, we see too much experimentation, invention, and imagination in Nature. Instead of saying that those things are unscientific fantasies of the human mind, many speculative thinkers make the opposite point. To arrive at DNA, life on Earth, and the human mind, the universe was self-aware and could understand what it was doing. Science is obligated to accept the simplest, most elegant explanation for things. It is far simpler to accept consciousness as a given than to come up with tortured schemes that become ever more complex by denying the central role of consciousness.

Creation without consciousness is like the fabled roomful of monkeys randomly striking keys on a typewriter until they wind up, millions of years later, writing the complete works of Shakespeare. One researcher actually arranged to have a random-number generator (an updated monkey) spew out letters to see if sensible words would emerge. They did, but it took countless tries to form even a simple phrase, and the unlikelihood of producing *Hamlet* was astronomical. (As a character, Hamlet speaks 1,495 lines, and if our computer-monkey got the last syllable wrong—writing "The rest is silent" instead of "The rest is silence"—it would have to repeat the whole random process from the start. Only thirty-six plays to go!) Human DNA is thousands of times more complex in its structure than the letters composing Shakespeare's writings. Rather than supposing that Nature had to go back to the beginning every time it randomly left out a gene splice, it's more reasonable to assume that the universe remembers the steps of evolution and can build upon them. In other words, it is self-aware, or conscious.

Spirituality, then, has viable arguments about how the universe came into existence, arguments that transcend Leonard's mathematical model because that model is insufficient. Math doesn't begin to explain why the ingredients of the early universe look suspiciously like the exact materials needed for conscious life. As the noted theoretical physicist Freeman Dyson writes, "Life may have succeeded against all odds in molding a universe to its purposes." For those who insist on the primacy of matter, there is even convincing material data for throwing randomness out the window. At the time of the Big Bang, the number of particles created was slightly more than the number of antiparticles. There were a billion plus one particles for every billion antiparticles. These particles and antiparticles instantaneously collided and annihilated each other, filling the universe with photons. Because of the initial tiny imbalance, however, there were excess particles left over after the annihilation, and this alone created what we know as the material world. What are the chances of that? About the same as the chances of blowing up a skyscraper with dynamite and finding a new skyscraper springing up from the leftover dust.

Leonard has offered even more intricate descriptions of the first few seconds after the Big Bang, but I want to stay with a simple concept. If all you care about is the data, then you and I, and all living species, along with the stars and galaxies in our universe, are the result of a freakishly small imbalance at the moment of creation. The physical universe had every likelihood of not emerging. But it did, and something else emerged along with it: an organizing force that shaped the roiling, chaotic infant cosmos without itself being visible.

In the absence of that shaping force, the odds against you and me appearing are too fantastically small to be credible. Physicists have added many other coincidences to the ones Leonard enumerates, but he minimizes the baffling state of affairs that has resulted: the universe's parts fit together with infinite and infinitesimal precision. No matter how small the scale or how large, the cosmos is seamlessly exact in a way that randomness cannot account for. *Something* must have caused this, and it must exist beyond the visible universe. Even by their own lights, materialists confront a transcendent realm, and throwing God out of that realm won't make it any less real.

Still, to arrive at a new creation story, there is no need to invoke God in a traditional sense (even though Einstein's awe and wonder are, according to him, completely necessary if someone wants to make great scientific discoveries). What is crucial for my side of the debate is that science has been forced to peer into the void that exists beyond time and space, opening the door for consciousness, creativity, evolution, orderliness, and wholeness as basic principles in Nature. As I will show, without these traits, the universe could not have produced DNA, life on Earth, the human mind, and civilization. Since they all exist, the cause for spirituality is far from lost. It's just beginning to assert itself.

3

Is the Universe Conscious?

DEEPAK

In an old Jewish punch line, God creates the world, sits back to view his handiwork, and says, "Let's hope it works." In science's creation myth, nothing creates the world or has any idea if it will work. The universe was mindless until the arrival of the human brain, which looked back at its own evolution and declared, "Nothing could be conscious but me. There is no awareness outside me or before me."

The curious thing is that physics, in proposing a universe where consciousness has no place for 13 billion years, undercuts its own foundation. The most advanced aspect of physics, quantum theory, tells us that a subatomic realm provides our best description of nature—the quantum field that holds reality together. But then physicists place this field outside ourselves: in other words, human consciousness knows itself, but the field isn't permitted to do the same. This exclusion forces science into some tortuous claims. For example, Stephen Hawking publicly declared his support for the existence of trillions upon trillions of other universes (the exact number being 1 followed by five hundred zeros). None of these alternate universes has ever been seen or proven. The need for other universes is to have enough spares to throw away, because if you hold, as Hawking does, that consciousness is the outcome of random physical processes, it takes a lot of failures before one lucky universe—ours—hits the jackpot.

Against this fantastical conception of trillions of leftover universes, I'd like to quote the *Bhagavad Gita,* at the moment when Lord Krishna wants to describe his divine nature: "I am the field and the knower of the field." In ten short words he marks out the spiritual side of the debate. There is a field that comprises all of creation, both visible and invisible, and it is imbued with a mind that knows itself. (Although physics defines "field" in a narrow technical way, the ancient usage simply means the ground of existence.) When they explored their own awareness, the great sages of ancient India discovered *"Aham Brahmasmi,"* which means "All that exists is within me," or in simplest terms, "I am the universe."

Aham Brahmasmi states something very basic: consciousness exists everywhere in Nature. If you reject this notion, the alternative is nearly absurd, because it turns consciousness into an accident, the chance result of DNA being boiled up in the chemical soup of the Earth's oceans two billion years ago. Then, through a chain of equally haphazard events, human intelligence evolved in order to look out at the cosmos and say, "I'm the only one who can think around here. Aren't I lucky?" (I was told by one physicist who became interested in a conscious universe that she was heckled at a conference by senior physicists, one of whom cried, "Go back and start doing good physics again." She noted that their younger colleagues looked interested but kept quiet.)

As we've seen, the weakest link in the current argument from science is randomness. Substitute a car factory for the visible universe. The factory's assembly line produces beautifully made machines, intricate and efficient, each design displaying invention and creativity. Yet when you go around to the back of the plant and look closely, you find a cloud of iron atoms, silica, and plastic polymers swirling mindlessly as they are sucked into the factory. Is it really credible that this cloud of matter and energy, plus an indeterminate amount of time, was enough to lead to a car, all on its own? That is science's current story about how the Big Bang led to the human brain. Incredibly, when asked if perhaps the Big Bang contained the potential for creativity and intelligence embedded in it, science's conventional answer is a resounding no. Chaos can produce those things, we are told, given enough time and trillions of random interactions.

Some scientists, uncomfortable with a blind creation, have tried to awaken the cosmos a little, and sometimes a lot. Sir James Jeans, an eminent British physicist in the first half of the twentieth century, mused, "The universe begins to look more like a great thought than a great machine." In our time Sir Roger Penrose, another renowned English physicist (and a frequent debater with Stephen Hawking), proposes that the seeds of consciousness are embedded in the universe at the finest level of Nature, the vanishing point of matter and energy (technically known as the Planck scale of space-time geometry).

Penrose speaks of mathematical truth, for example, as being a Platonic value, named after the Greek philosopher Plato, who proposed that every human quality was born from a universal quality—for example, love is a Platonic value because it is inherent in creation, not something invented by humans to describe their emotions. We feel love because we are part of creation. Penrose relies on the fact that all of science rests upon mathematics, but he sees math as more than just numbers to be crunched. To someone who really understands it, mathematics expresses values that reflect the cosmos, including orderliness, balance, harmony, logic, and abstract beauty. You can't strip the numbers out and leave these other values behind.

Every physicist agrees to the preeminence of mathematics, so it's hard to see how science can get away with rejecting the qualities that go with mathematical reasoning. In other words, if you are looking for truth, doesn't truth have to be part of the setup of your mind? Otherwise, how would you know what to look for? Once you have embedded harmony and logic in the fabric of the cosmos, you have a much harder time excluding consciousness. Spirituality takes the next logical step: everything we experience occurs in consciousness; therefore, there is no reality "out there," divorced from consciousness. Penrose won't go this far, since he is on record as declaring that he abhors the notion of a subjective universe. But the beauty of invoking cosmic consciousness is that we can do away with the war between subjective and objective. In the universe's precreated state the potential for both existed, as seeds in the womb.

Other thinkers have taken a deep breath and let the whole thing in. Instead of isolating the human mind from the field of creation—like a hungry child with his nose pressed against the bakery window—some scientists choose to break down the barrier between the universe and ourselves. The late John Wheeler of Princeton held that the visible universe could come into existence only if someone observed it, and without such an observer, there would be no universe. Minus the participation of an observer, the universe would still be in a state of pure potential. When we gaze at the stars, is that what makes them appear?

Cries of "solipsism" may fill the air, but it isn't necessary to say that the universe waited for human beings before it came into existence. The observer could be God. (Now cries of "faith" and "superstition" fill the air.) But we don't need God, either. All we need is a universe that contains consciousness as an inseparable aspect of itself. Once you grant that, then any and all observers—divine, human, or any other kind—are expressions of self-awareness. They share the same status; each is a participant in creation. The great opportunity for spirituality to rescue science from a blind creation is that it allows conscious beings (us) to participate in a conscious universe.

But what does "participate" really mean? When a physicist like Wheeler reasons that in the beginning there were only probabilities, he is talking about a well-known concept in physics, the collapse of the wave function. An elementary particle like a photon doesn't simply exist in time and space like a shiny little ball hanging from the Christmas tree of the cosmos. Photons carry light in tiny packets, but they also behave like waves. Waves extend in all directions, forming the electromagnetic field that spans the universe. There is a probability of finding a photon anywhere in the field, but as soon as you detect one somewhere, you don't need a probability. The very act of observing has transformed the wave into a particle.

To me, the fact that a particle can exist in an invisible state has immense implications (some of them unacceptable to workaday physicists), and the most important one for spirituality is this: Before the Big Bang the state of the universe contained all possibilities. Everything that does exist—or ever could—derives from that original state. In everyday life this doesn't seem like a statement with practical implications, but it is. Consider your use of English. Before you pick any word to say, such as "elephant," it is only one possibility. You may or may not pick it. You might pick "pachyderm" instead, which exists as another possibility. But once you do pick a word, an event has occurred in the physical universe, and the possibilities that you might have chosen in that moment (but did not) have remained in the state of pure potential.

The strange thing, so far as logic goes, is that no matter how many possibilities turn into reality, an infinite number still remain. The visible universe is only a tiny bit of what *could* exist. All the possibilities that didn't collapse are still there, just as real as the ones that did. Consciousness works the same way. When you pick the word "elephant," your vocabulary still contains thousands of words that you didn't use. The unused words aren't destroyed or forgotten; they remain as possibilities. Here we are, you and I, participating in genesis right now, and in every moment. Lord Krishna says about the process: "Curving back upon myself, I create again and again."

If the field contains everything that could possibly be, we cannot exclude consciousness or human values. Here spirituality can enrich science. Physics blithely dismisses the all-too-human need for the cosmos to be a meaningful home, a nurturing place for love, truth, compassion, hope, morality, beauty, and every other value once ascribed to God. Since these qualities have no mathematical validity, science feels free to banish them. But in reality we pluck these values out of the universe's infinite possibilities, just as we pluck words out of our vocabulary.

Even though Roger Penrose—and almost every other senior faculty member in the field of science—abhors the notion of a subjective universe, it doesn't have to be thrust upon him. Spirituality isn't about substituting subjectivity for objectivity. Some paranoid schizophrenics are convinced that the world will disappear if they fall asleep, so they try to remain awake twenty-four hours a day in service to humanity.

But the Buddha and the Vedic sages aren't saying that's necessary. They are saying that a primal state exists that embraces both subjectivity and objectivity, a premise that is totally consistent with quantum reality. Once the wave function collapses, there is a subject-object split: now "I" am looking at "a thing." But before the subject-object split, reality is one infinite entity. It must be that way if all possibilities are contained in it.

There is much more to say about how the human mind and the cosmic mind are linked. Once you admit that the universe might be self-aware, there is suddenly no mystery as to why humans are intelligent, creative, and conscious. It's in the air we breathe; it's in the neighborhood where we grew up. In fact, the domain of infinite possibilities is the closest thing to us all the time. As the mystic Persian poet Rumi put it, "Look at these worlds spinning out of nothingness. That is within your power."

Whatever the universe contains, including us, must exist in potential first. The source keeps tabs on creation because it is actually keeping tabs on itself. This is the role that consciousness plays, and by not recognizing it science blindfolds itself. From the spiritual viewpoint, the probability waves of quantum physics inhabit the same dimension as the mind of God, which the greatest scientists throughout history have always hoped to fathom.

LEONARD

Friedrich Nietzsche wrote, "Formerly one sought the feeling of the grandeur of man by pointing to his divine origin: this has now become a forbidden way, for at its portal stands the ape, together with other gruesome beasts, grinning knowingly as if to say: no further in this direction!" That was in 1881, ten years after Darwin wrote *The Descent of Man,* in which he proposed that even the noblest features of human beings were the result of the same processes of randomness and natural selection that produce the quack of the duck and the slither of the snake. Darwin's theory of evolution has grated on people ever since he made it public with *On the Origin of Species.* In one early encounter, according to legend, Samuel Wilberforce, bishop of Oxford, asked T. H. Huxley, a staunch Darwin supporter, if it was "through his grandfather or grandmother that he claimed his descent from a monkey?" Huxley is said to have answered essentially that he would not be ashamed to have descended from an ape, but would be ashamed to be connected to a man who argued like Wilberforce. Today, ironically, physicist Stephen Hawking, a man who has done much to banish the need for a divine origin from our understanding of creation, has his Cambridge University office on none other than Wilberforce Road. The détente is not universal. Plenty of scholars today, religious and otherwise, feel a need to attribute the grandeur of humanity to our special connection with the divine.

Deepak calls the scientific explanation of how we got here "science's creation myth." In employing such terminology he equates the careful observation and theoretical work of science with the legends and speculations of ancient civilizations, some of which form the basis of his own beliefs. But that anything-goes approach is not a productive path to truth. Deepak considers distasteful a universe in which consciousness did not exist before the arrival of human beings. He prefers a rosy picture of a universal consciousness that has been present ever since creation. However, if we don't subscribe to the anything-goes approach, the issue is not whether a conscious universe is preferable, but whether a conscious universe is real. Wish fulfillment should not shape our worldview.

What would it mean for the universe to be conscious? Scientists have a difficult time attaching a precise definition to "consciousness," though we all have a rough idea of what the term means. One quality always included in consciousness is self-awareness. In contrast, cerebral processes that are automatic, beyond willful control, and of which we are not aware are considered unconscious. Experiments with mirrors seem to indicate that chimpanzees and orangutans, and even magpies, do have some self-awareness, in that they recognize the image in the mirror as themselves. Nematodes and fruit flies presumably don't, so self-awareness draws a certain line among the species. Still, self-awareness alone is a crude classifier, and most of us would like to think that the ones handing out the bananas have at least a higher level of consciousness than those receiving them, so consciousness probably comes in degrees.

Consciousness also varies with the state of our mind. For example, we all have nonconscious periods, occurring in what is called slow-wave, or deep, sleep. If you ask normal awake people to describe what they were thinking or experiencing just before you asked them, they can tell you. This is also true if you wake someone during rapid eye movement, or dreaming, sleep, though the dream may quickly fade from memory. But if you awaken people during deep sleep they will have nothing to report. Their minds will be blank notebooks. Indeed, recordings of neural function during deep sleep show only activity associated with automatic, unconscious cerebral processes.

Another complication in defining consciousness is that our conscious and unconscious minds are coupled systems. There has been much recent research into the effect of the unconscious on what we think of as conscious social behavior and decision making. But the most vivid example of conscious actions based on information the conscious mind is unaware of comes from a phenomenon called "blindsight." Blindsight results from damage to a part of the brain called the primary visual cortex. As a consequence, people afflicted with blindsight fail to consciously see anything in all or part of their field of vision, a situation that can be confirmed through brain scans. However, we know that in those with blindsight, images picked up by the eye are nevertheless transmitted to the brain, where, without ever reaching the level of conscious experience, they influence conscious behavior. Thus, people with blindsight can reach out and touch objects, catch objects you toss to them, distinguish smiling from angry faces, and even, in one case, navigate an obstacle course, all without being aware of having seen anything.

We infer the consciousness of other humans, or animals, by interacting with them. But we can't hold up a mirror to the universe to see if it preens. If the universe is conscious, how can we tell? It would be like a cell of the stomach lining knowing that when it is inflamed, the individual it is part of feels the ache. It is tempting to believe that consciousness (preferably a loving and compassionate consciousness) plays a role in the physical universe. In fact, natural philosophers for centuries believed that physical laws were analogous to human laws, and that objects in the universe consciously obeyed those laws because they wished to avoid the punishment of the gods. Even as late as the seventeenth century, the great astronomer and physicist Johannes Kepler believed that planets followed laws of motion that were grasped by their "minds." But that idea did not lead to any testable consequences, so science abandoned it. The idea of universal consciousness is equally barren, so it is best to abandon that idea, too.

Deepak says that science displays a stubborn resistance to other ways of regarding the cosmos, but the "other ways" science resists are merely ways for which there is no supporting evidence. Deepak laments that "we have assigned cosmology to specialists the way we assign genes to geneticists." But I'm sure Deepak would agree that there are some enterprises that benefit from the work of specialists and some that don't. For example, we probably both think that pretty much anyone can make peanut butter and jelly sandwiches, but if one or both of us had to have heart surgery, we would certainly want a top-notch cardiac surgeon to do the job. Where Deepak and I seem to disagree is that I consider cosmology more like surgery, and he considers it to be sandwich making.

Deepak also warns that you can't say, "Keep out; you don't know enough math." I agree that people should be free to discuss whatever intellectual issues interest them, but we shouldn't confuse discussing and learning about a topic with creating a meaningful theory about it. Anyone can speculate whether or not the sun can go on shining like this forever, but it takes mathematics to give the speculation substance, and to fill in details such as that in seven billion years the sun will grow 250 times larger and swallow up the inner planets.

I embrace the preeminence of mathematics in science. It allows scientists to calculate numbers and to determine the logical consequences of scientific statements. It also helps us make precise and unambiguous definitions. It is easy to convince oneself of dubious ideas if the arguments one uses to support those ideas are built around words with wrong, vague, or multiple meanings. In fact, it is a theorem in mathematics that if you accept a false statement as true, you can use it to show that any other false statement is true. So precision of language is important, and the tools of mathematics are a great help in ensuring that concepts are precisely defined.

I agree with Deepak that mathematics is more than numbers to be crunched. I agree that mathematics is also about orderliness, balance, harmony, logic, and abstract beauty (though it is also about randomness and disorder). Scientists do not reject Deepak's values. We do not banish love, truth, compassion, hope, morality, and beauty from our thinking, but we do banish them from our theories. Would Deepak prefer that our equations say that the sun gets a fuzzy feeling when a pretty comet flies past? Should physicists punctuate their mathematics with theorems about the emotional state of a nebula? Can we appeal to the creativity of the universe to prove the Big Bang? Subjectivity is an important part of human experience, but it doesn't mean we must incorporate love into our theory of the orbit of Mercury, or universal consciousness into our theory of the physical universe.

Lord Krishna might have said "I am the field and the knower of the field," but it is a good bet he never designed a radio. There is plenty of room in human experience for Lord Krishna's teachings, but that doesn't mean one gains by incorporating them into science. Physics proposes a universe in which consciousness has a place within human beings—and within other animals on Earth and possibly on other planets—but that is where nature seems to draw the line. Stephen Hawking might theorize about trillions upon trillions of other universes, but he doesn't foresee them theorizing about him. And until our observations of the cosmos indicate otherwise, few scientists are likely to consider the universe a conscious entity.

4

Is the Universe Evolving?

DEEPAK

Evolution is the club that science wielded to beat religion into the dust, and whenever religious ideas threaten to take on new life, science rushes in to smash them down again. These ideas include, first and foremost, the perfection of God. According to religion, the deity didn't need to get smarter, because God is omniscient. He (or she) didn't need to expand into new places, because God is omnipresent, or to increase in power, because God is omnipotent. Having declared the creator perfect, religion couldn't call God's creation imperfect; therefore, the universe didn't need to evolve, either. But the rise of intelligent life from primitive life-forms is undeniable. Physics has proven that the universe expands, and that energy gathers into vast clumps known as stars and galaxies that are more organized than interstellar dust. The defeat of perfectionism seems totally justified. We live in an evolving universe.

Spirituality therefore cannot get back into the game on religious terms. It has to add something new to the concept of an evolving universe. I think it can. If consciousness underlies everything in Nature, it is the force that directs evolution. If not, then evolution becomes, along with everything else, the result of blind random activity. Physics has chosen the second assumption, which has led it to some glaringly false conclusions.

First, science focuses on physical expansion as the basic foundation of evolution. At the instant of the Big Bang the known universe was billions of times smaller than the period at the end of this sentence.

Now it spans billions of light-years. But that expansion isn't evolution, any more than blowing up a house with dynamite is. The house certainly expands when you blow it up, scattering its fragments in all directions, much like the Big Bang did for the universe, when an unimaginable blast of energy scattered elementary particles in all directions. Yet behind the mask of matter, something more mysterious was happening.

To get at the mystery, let's follow the path a hydrogen atom might take over the thirteen billion years or so following its creation. First it drifts out into space in a completely disorganized, random fashion, bouncing around like an infinitesimal feather on the cosmic wind. Some atoms keep on doing this until they form clouds of interstellar dust. But this atom falls into a stronger gravitational field and becomes a building block for a star, which takes primitive atoms like hydrogen and helium and transforms them into heavier, more complex elements. Through a series of nuclear reactions our particular hydrogen atom becomes part of the element known as iron, the heaviest metal formed inside stars.

The life span of this star comes to an end in the dramatic death throe known as a supernova, an enormous explosion that scatters iron atoms throughout the nearby regions of the cosmos. Our original hydrogen atom no longer exists as such, but its component parts are being drawn toward another star, hundreds of times smaller: the sun.

By this point in the history of the universe, the sun has already thrown off enough matter during its birth pangs that rings of dust have settled into orbit around it. This dust is clumping into planets, and our iron atom, pulled in by gravity, joins the planet Earth. At its core, the Earth is thought to be up to 70 percent molten iron, but our atom arrives late enough to settle onto the surface of the planet, which is around 10 percent iron.

Ten billion years have now passed. Many iron atoms have undergone random interactions with various chemicals, but ours is still intact. More time passes. It finds itself drawn into a spinach leaf, which gets eaten by a human being. Then our iron atom becomes part of a molecule thousands of times more complex than itself, a molecule that has the ability to pick up oxygen and throw it off at will: hemoglobin. Hemoglobin's ability to perform this trick turns out to be crucial, because another molecule, this one millions of times more complex, has managed to create life. It is known as DNA, and around itself DNA is gathering the building blocks of life, known as organic chemicals, of which hemoglobin is one of the most necessary, since without it, animals cannot convert oxygen into cells.

In our story, one primal hydrogen atom has undergone incredible transformations to get to the point where it can contribute to life on Earth, and every step of the way involves evolution. Since all the iron on Earth was once part of a supernova (plus some iron deposited when meteorites collided with the early planet), the journey from the Big Bang can be observed and measured. Yet our iron atom has still another transformation to undergo. It has entered the bloodstream of a human being—you or me, perhaps—to become part of a sentient, thinking creature, one that is capable of looking back on its own evolution. In fact, this sentient creature created the notion of evolution in order to explain itself to itself. A primal atom has somehow become thoughtful.

I've taken the time to follow a single atom for 13.7 billion years because the steps it took to arrive in my body or yours, allowing me to write this sentence and you to read it, encompass the invisible qualities that spirituality is all about: creativity, quantum leaps of transformation, the emergence of unexpected properties, and overall, an enormous display of intelligence. As evolved creatures, we attribute all these qualities to ourselves. So where did they come from? Physics claims that they came from random physical processes, but that answer makes no sense. At every single step of its journey, our hydrogen atom resisted randomness. It became more complex; it contributed to increased energy; finally, it made the leap to human intelligence. The iron that allows you and me to be alive and sentient is no different from the iron in a rusty sewer pipe, or in interstellar dust. Yet evolution had a different fate in mind for our atom, and spirituality claims that its fate was directed by consciousness.

Consciousness-directed evolution isn't the same as invoking a creator God. Instead, it introduces a property inherent in the cosmos: self-awareness. The beauty of this property is that it can include randomness; there is no need for an either/or choice. If you take a highly ordered molecule like hemoglobin, which contains thousands of perfectly arranged atoms, like thousands of dewdrops on a spider web, you can examine it at finer and finer levels. As you get to the quantum level, atoms are considered clouds of probability. The dewdrops have evaporated into a mist. Because science is reductionist, it claims that random electrons emerging from probability waves provide the ultimate explanation for the visible universe, based on chance but guided by basic forces like electromagnetism.

In spiritual terms, this is a topsy-turvy explanation. It's very hard to get to life on Earth starting from total chaos, much more difficult than shaking a beaker of stem cells, walking away, and then coming back to find Leonardo da Vinci. Why not explain creation by what it achieves, instead of by what it can be broken down into? The Great Pyramid of Cheops can be examined as a heap of different kinds of dust, but that doesn't explain it, any more than breaking the human body down into subatomic particles explains who we are. As the noted English physicist David Bohm put it, "In some sense man is a microcosm of the universe; therefore what man is, is a clue to the universe." The music of Bach can be broken down into sound waves, but once you arrive at this raw data, you lose Bach. His genius has been reduced to the same level of information as a clap of thunder or a rumbling earthquake.

The great flaw of reductionism is that when it pushes out the invisible aspects of creation, it thinks it has improved our understanding. Turning around and saying that data is actually *better* than the messy, ever-changing thing we call experience is totally wrongheaded. As the great quantum pioneer Niels Bohr put it, "Everything we call real is made of things that cannot be regarded as real." To someone who insists that solid objects are the only real things in the universe, this is a fatal blow.

Evolution stops short of being God. Rather, it's the tendency for the universe to unfold along steps of increasing intelligence. A huge amount of wiggle room is left for experimentation, side trips, detours, and sudden leaps. This fizzy, uncertain, yeasty reality has been with us since time began.

Spirituality will win the struggle for the future by restoring consciousness to evolution. The next step depends on us. Human beings must break away from materialism if we want to keep evolving. As a species we alone can transcend biology. In fact, the process is already well under way. We have crossed the crucial divide. Science is proof that we have taken conscious control of our own evolution, and so is spirituality. The guiding hand has let go, allowing us more and more freedom. When we accept it, our participation in the universe will take a quantum leap: we will fully become cocreators of reality. Evolution isn't the whole of the mind of God. It is only one aspect, the one we are about to claim as our own.

LEONARD

A quick way to turn science into science fiction is to play with the meaning of its terms. When an astronomer says the sky is alive with stars, she doesn't mean you can trade recipes with it. So if we say, quite catchily, that "evolution is the club that science wielded to beat religion into the dust," and then ask if the universe is evolving, we'd better get straight what we mean by "evolution." In common parlance evolution is "any process of formation or progressive change." In biology (the field that ostensibly used evolution to club religion to death), it means "a process that produces change in the gene pool of a group—via mechanisms such as mutation and natural selection—that is heritable from one generation to the next." There are two differences in these definitions. First, the scientific meaning of evolution refers to a specific change, an alteration in the genes of a group of organisms. Second, it specifies the mechanism of change. Natural selection is a process in which organisms better able to cope with their environment tend to have more offspring, which creates a new generation that on average has more traits favorable to survival and reproduction than the last.

Natural selection is what makes evolution more than just a random process. If you ignore it, you can indeed make the theory of evolution appear absurd and far-fetched. For example, Deepak writes that "creation without consciousness is like the fabled roomful of monkeys randomly striking keys on a typewriter until they wind up, millions of years later, writing the complete works of Shakespeare." He tells about a researcher who arranged to have "a random-number generator (an updated monkey) spew out letters to see if sensible words would emerge." Since it took countless tries to form even a simple phrase, and since human DNA is thousands of times more complex in its structure than the letters constituting Shakespeare's works, Deepak concludes that the theory of evolution could not possibly account for the structure of our DNA. That random-typing experiment is typical of the kind of misleading arguments that arise when you ignore natural selection. Richard Dawkins addressed it in his book *The Blind Watchmaker.* He described a computer program he wrote, which included a mechanism analogous to natural selection. Setting it in motion, he waited to see how long it would take for the program to arrive at Shakespeare's phrase "Methinks it is like a weasel" through random typing in a manner that mimics evolution. In the purely random model Deepak described, the chance of typing the entire phrase correctly is one in ten thousand billion billion billion billion, so a computer could generate random string after random string in this manner until the sun burns out and still never hit upon the target phrase. But by incorporating natural selection into his random-typing program, Dawkins showed that the phrase could be produced in just forty-three generations—a mere moment or two on a decent computer. That is the magnitude of the error that can arise if one is not careful about the precise definition of concepts in science!

One cannot apply the Darwinian concept of evolution to the universe as a whole, because concepts like heredity and natural selection—by which individuals less able to survive their environment die out and the gene pools of those that are more fit prevail—make no sense in that context. A cloud that changes shape from an elephant to the face of Jesus cannot be thought of as evolving according to the biological meaning of the word. Nor can a spinning cloud of interstellar dust and gas that flattens and condenses into a star and planets. Such a system can be said to be evolving in the sense of everyday language, and physicists might on occasion use the word in that sense, but its progression has nothing to do with the theory of evolution that "beat religion into the dust." So is the universe evolving? The universe is undergoing progressive change, but that is not evolution in the sense that Darwin made his name on.

Having locked Darwin in the basement for now, we can deal with the real issue. Is the universe evolving, in the colloquial sense, toward greater complexity and intelligence? And, if so, is there evidence that the trend is the result of a guiding force such as consciousness? Is the march of the cosmos an evolution toward something higher? Have scientists overlooked the existence of meaningful progressive change in this universe that is our home?

The answer is, again, no. In later chapters we'll see that even biological evolution does not have any "innate" drive toward intelligence and complexity, but as regards the physical universe, the opposite is in fact true: the universe, I am sorry to say, is heading toward a simple and lifeless end.

Why is that the future of the universe? As I explained earlier, the universe is expanding. That expansion will continue at an ever-increasing rate. As that happens, the matter and energy within the universe will grow ever colder and more dilute. Distant galaxies will eventually move so far away that we will no longer be able to detect them. Eventually all that will remain in our observable universe will be our local group of galaxies, bound to us, if weakly, by gravity. Astronomers living then could conclude that our galaxy, and perhaps a few neighbors, are all that there is in the universe, or ever was. They might have no way of knowing the rich history that preceded.

Sadly, those isolated worlds, too, will eventually end, for stars burn out. They can end their life cycle in different ways: they can collapse into black holes or neutron stars; they can fade like glowing embers, becoming a type of star called a white dwarf; or they can explode as supernovas. In the last case, new stars and solar systems can form from interstellar gas and debris, leading to new life, but with time supernova explosions will become rarer, and eventually cease, and the reservoir of interstellar gas will become dilute and "dry up." When that happens the universe will consist of just the corpses of dead stars:

white dwarfs, black holes (which will eventually "evaporate"), and neutron stars. None of these can sustain life, so the universe will then be utterly dead. And if physicists who believe the proton is unstable are correct, even these corpses will break up and dissipate, leaving a universe that is nothing more than a thin gas of particles within a vast void. This may seem to be a depressing picture, but as my mother told me when I was three and learned that people die—don't worry, the death of the universe is a long time off. Perhaps as many as 10,000,000,000,000,000,000,000,000,000,000,000,000,000 years.

If Deepak is right that the universe is purposefully becoming ever more complex, then the picture I just painted is wrong, and some of the most fundamental and well-tested principles of physics are also wrong. But if this picture is correct, if the development of the universe is not purposeful, and not evolving toward ever greater complexity, then how do we interpret Deepak's story of the lone hydrogen nucleus, born in the early universe, improving its lot by becoming part of that princely metal, iron, and eventually making its way into a conscious human being? How could such an unlikely event happen? Could it really occur through random processes?

Beautiful and ordered objects arise from the purposeless laws of nature all the time, from rainbows to snowflakes. But human beings are predisposed to search for patterns and, once we've found them, to assume they are born of good cause. We don't need to look to cosmology to be fooled by randomness. In *The Drunkard's Walk* I wrote about the case of a mutual fund manager named William Miller. He became famous for running a fund that outperformed the Standard and Poor's index for fifteen years straight. Thousands of mutual fund managers over several decades were all trying to accomplish that feat, but only one manager did it. Even to many who think that stock picking is of marginal value at best, it seemed that feat could have been accomplished only through a relentlessly brilliant knack for anticipating the futures of individual stocks, and investing accordingly. But the mathematics of probability yields a surprising result: if you replace those thousands of managers with gamblers who simply flip a coin once each year with the goal that it come up heads, you'll find that the chances are very high that one of those gamblers, too, will have a streak of fifteen or more successful years. William Miller's much-heralded feat, it turns out, could indeed have resulted from randomness alone.

The story of the "evolving" hydrogen atom is analogous: our awe in the face of the unlikeliness of a rare feat can be neutralized by knowledge of the great number of opportunities for such a feat to be accomplished. Supernovas, for example, are extremely unlikely events. If you pick a typical galaxy of, say, a hundred billion stars, you'd have to stare at it on average for an entire century before you'd see one of those stars explode. Yet if you hold your arm out and block a patch of sky with your thumbnail, there are so many galaxies in that patch that, with a sufficiently powerful telescope, you'd see ten supernovas each night. Rare events happen all the time.

In the case of the proton, there are roughly 10^{80} bouncing around in the observable universe, only a very tiny fraction of which end up a cog in some life-form. In fact, on Earth there are about 10^{42} protons in the biomass, so even if we assume that every star in the observable universe has its own life-friendly Earth—and probably few actually do—we find that for every proton that stumbles its way into a living organism, there are at least 10,000,000,000,000,000 protons stumbling around that don't. Just as once in a blue moon a coin flipper can achieve fifteen heads in a row, on very rare occasions, without the intervention of any conscious force, so too can a proton end up, not in a star, or in interstellar space, but inside a living thing. Science doesn't say that nature shook a beaker of stem cells, walked away, and came back to find Leonardo da Vinci. It says she sent matter into a billion trillion star systems, let it brew for 13.7 billion years, and *then* produced a Leonardo da Vinci. The former is indeed far-fetched; the latter is the beautiful consequence of the unguided and purposeless forces of nature.

If scientists describe the universe through laws that act without purpose, it's not because we oppose an intentional universe; it's because we don't appear to live in one. It can be inspiring to believe the universe is evolving toward greater complexity and intelligence under the guidance of a universal consciousness. But for scientists, such musings are not where the investigation ends; they are where it begins. Deepak attacks science's use of reductionism as an approach to understanding the universe, but scientists are not wed to a single method. When a phenomenon can be easily explained by reducing it to its simpler elements, scientists do that. When it cannot, when it depends on the collective interactions of a great number of components, we recognize that, too. Thus, when chemists study the properties of water, they analyze its molecular components. But when oceanographers study waves, they are not interested in dealing with the finer constituents of the water. Science has theories of water molecules, and theories of water waves, and having one does not exclude having the other. The end of an investigation comes when, regardless of an idea's attractiveness, we are able to find evidence to prove it either right or wrong.

If the universe evolved through physical law and had no guiding purpose, no consciousness, does that negate the value of humankind, or make our lives meaningless? Is the scientific view a heartless view of life? My mother, now almost ninety, told me once of a cold day when she was about seventeen, and the war was raging in Europe. Her town in Poland was occupied by the Nazis, and on this day one of those Nazis told a few dozen of the town's Jews, including my mother, to line up in a row and kneel in the snow. He walked the row and, every few steps, leaned down, put his gun to someone's head, and fired. The spiritual view says that my mother's survival was not random. It says my mother was passed over for a reason. Does this not imply that there was also a cosmic reason that those *not* passed over were slaughtered? Since most of the members of my parents' families *were* killed during the Holocaust, to me it is this "spiritual" explanation that feels cold and heartless.

Science offers a different view: The human animal evolved to have the capacity for both good and evil, and it does plenty of both, but there is no hidden hand of universal purpose or consciousness behind what we do, only our own consciousness, our own purpose. Each of us chooses love or hate; we give and we take; we leave our own imprint on our family, our friends, and society. We don't need an eternal and conscious universe to give our lives meaning. Our lives are as meaningful as we make them.

5

What Is the Nature of Time?

LEONARD

A few years ago researchers interested in the subjective perception of time arranged to have volunteers harnessed to a platform, raised a hundred feet into the air, and dropped into a net at an amusement park in Dallas, Texas. Before any of the twenty participants had a turn, they observed someone else being dropped. After this preview they were asked to shut their eyes and imagine the fall. They were told to press a button at the moment when they pictured the person beginning to drop, and again when they pictured the person landing. Then each subject took the plunge themselves. Afterward, they were asked to imagine their own fall and as before to press a button at the beginning and the end. The subjects' mental playbacks of their own experience lasted significantly longer than both their imagined experience of others, and the actual experience. The researchers had expected this because people who have endured brief dangerous events, such as violent attacks and car accidents, often report that the events seem to have occurred in slow motion. But our memory of an event depends on two neural systems—that governing our perception of the event, and that governing its recording and recall from our memory. So one might ask, Do we really perceive dangerous events in slow motion, or do we just remember them that way? Do we have a single sense of time that becomes distorted, or does the clock of our perception of the event run at its usual pace, but the clock of our memory of the event slow down?

To investigate that issue the subjects were given a wristwatch that flashed random numbers and told to read the digits during their fall. The catch was that the digits flashed just a bit too quickly for them to make out—that is, too fast to make out in ordinary circumstances. If the stretching of time that affects the memory of such events also affects perception, the falling subjects would see the numbers as flashing more slowly, and be able to read them. But the subjects couldn't read the numbers. Their memories recorded the events in slow motion but their perception clock was unaltered.

Perception and memory clocks are not our only measure of time. We seem to have many internal clocks, underpinned by different neural mechanisms. Much of our feeling for time comes from the clocks built into our bodies and visible in our environment. The principal clock in our environment, the rhythm of day and night, light and dark, is intimately connected to at least one clock in our bodies, the circadian rhythm. Living things—even unicellular organisms—have this biological rhythm that runs on a sleeping and waking cycle of about one day. In many animals this is governed by a biochemical process in which certain proteins accumulate, enter cell nuclei, degrade, and cycle back to their original state. The process is more complex in humans, and takes place in a part of our brain called the hypothalamus. In all animals the twenty-four-hour clock is only approximate. Human beings living in total darkness will have sleep/waking cycles lasting about twenty-five hours, while mice and fruit flies kept in darkness have cycles that are somewhat less than twenty-four. But under normal conditions, these biological clocks are reset each day, in humans when photoreceptors in the eyes and skin cells pick up light from the sun. Animals have other built-in bodily rhythms that run on much shorter cycles, such as the in and out of respiration and the pumping of our hearts, as well as certain wave patterns that occur in our brains. It is through all these internal clocks that we feel the passage of time.

The multiplicity of biological clocks leads to some interesting illusions—for example, in one experiment subjects were fooled into thinking that a flash of light preceded their pressing a key when actually it came afterward. Biologists and neuroscientists are interested in understanding the subjective aspects of our sense of time, and the physical, chemical, and biological mechanisms that produce them, and indeed these are fascinating topics. But although your own memory clock might slow when you're tossed off a platform, for the rest of the universe it is business as usual. And so physicists, unlike biologists and neuroscientists, or saints and sages, ponder time's mysteries from a less personal standpoint.

The starting point for physicists is to examine what we mean by time. Human language excels at capturing human feelings, but we shouldn't let our language define our concept of reality. If you haven't thought much about it, time is hard to define. It is an abstract concept derived and distilled from our experience. We describe the motion of projectiles and planets employing time, but time isn't a material object. One can think of time as one thinks of space, as a coordinate that enables us to label events. The event of the opening of the heliport atop the World Trade Center has the coordinates 40 degrees 43 minutes north latitude, 74 degrees 1 minute west longitude, 1,350 feet above ground level, and the year 1972. From this point of view we can consider the universe as a four-dimensional space akin to the three-dimensional space we see around us. But time not only labels the moments events take place and orders them; it also assigns events a duration.

One of the first clocks used in physics, at least according to legend, was the pulse of Galileo, who used that rhythm to time the swing of a chandelier in the cathedral of Pisa. Today we use more reliable clocks, like the natural oscillations of atoms. For example, when an atom jumps from a higher energy state to a lower energy state, radiation is emitted. That radiation oscillates with a frequency determined by the difference in energy between the states. The radiation corresponding to the transition between two particular energy levels of the cesium-133 atom passes through exactly 9,192,631,770 cycles each second. I can say "exactly" here with confidence because since 1967 that has been, according to the International System of Units, the definition of a second. And so if we say the crystal in a quartz watch vibrates 32,768 times per second, we mean that if we started counting the oscillations of the crystal and the radiation simultaneously, at the precise moment the cesium radiation had gone through 9,192,631,770 cycles, the quartz crystal would be reaching its 32,768th vibration. This highlights an important related concept that is crucial to the definition of time as duration: the concept of synchrony. We measure the time one process takes by comparing it to some other standard process—like the ticks of a stopwatch—that has a concurrent beginning and ending.

This nice intuitive picture of time works well in everyday life, but between 1905 and 1916, Albert Einstein showed that it is only an approximation of the way nature really works. The approximation is perfectly fine if you don't measure time too accurately, and you consider objects that are moving much slower than the speed of light, and that are in gravitational fields not much stronger than those we experience on Earth. But in truth, Einstein showed, those concepts upon which our ideas of clocks are based, especially synchrony, and even the fixed order of events, depend on the state of the observer—by which he did not mean the emotional state.

The fact that two events one perceives as simultaneous can, from the perspective of another observer, occur at different times, probably sounds somewhere between strange and wrong. It might help to look at the same effect with regard to space. Suppose a person standing in the aisle of an airplane bounces a ball on the floor. That passenger would report that the ball hit the floor at the same spot, over and over again. To an observer on the ground, however, the ball would not be returning to the same spot, but rather tracing a line across the sky at over 500 miles per hour. Both observers are right, from their own perspective. Analogously, different observers may disagree about whether events happen at the same time, and if the observers are moving fast enough relative to each other, that disparity can be noticeable. This is an important point for our later discussion of the nature of reality, so we'll come back to it then.

The inability of moving observers to agree on simultaneity means that clocks can disagree, and that different observers can disagree on the duration of events. The referees working for the *Guinness World Records 2010* watched as the world's fastest hot dog eater downed sixty-six hot dogs in twelve minutes, but observers flying past at great speed would have measured the feast as having taken much longer. According to relativity, each clock measures its own local flow of time, and observers who are moving relative to each other, or are experiencing differing gravitational fields, will in general find that their clocks do not agree.

One can think of a clock as a kind of odometer for time. An odometer measures the distance you travel in journeying from one event to another, while a clock measures the duration that elapses between events. The distance an odometer measures depends on both the difference in the spatial coordinates of the two events—such as their latitude and longitude—*and* the path the odometer takes to get from one event to the other. According to relativity, the time a clock will measure between events also depends on the path the clock takes between the events. For example, suppose two fifteen-year-old twins watched the World Trade Center dedication in 1972, after which one was snatched up by aliens and whisked off on a very fast rocket ride, perhaps even passing near (but not too near) the powerful gravitational field of a black hole. If the abducted twin was returned to Earth and reunited with her sister at the dedication of a World Trade Center Memorial in 2013, the Earthbound sister would be fifty-six while the abducted twin might be just sixteen. Between the abduction and the reunion the odometer of the Earthbound twin would have registered many miles, and her clock an elapsed time of forty-one years. Her sister's odometer would have registered many more miles—but her clock, perhaps only one year—between the same two events. Einstein showed that there is no contradiction in this; it is just the way time works. This effect was confirmed by experiment in 1971, in which a very accurate atomic clock was flown around the world and compared to an identical earthbound clock. The effect of the clock's motion at those relatively slow speeds, however, amounted to a difference of only about 180 billionths of a second per circuit.

Since an hour spent in a moonlight stroll with a loved one does not feel the same as an hour spent justifying your work to a nasty boss, we are lucky to have our reliable cesium atoms, whose light will pass through 33,093,474,372,000 cycles each hour regardless of our state of mind. Though the biologist, the neuroscientist, and the physicist will all conclude that time depends on the observer, they mean it in different ways. To a physicist, time depends only on motion and gravity, and we have mathematical formulas that account for the relevant factors. These allow physicists to translate back and forth among different observers' clocks, without any bias that might arise from the observers' feelings entering into the physicists' formulas.

When we humans slow down to smell the roses, the betadamascenone molecules that carry the smell continue their motion unaffected by our subjective feelings. But when the Earth exerts its gravitational pull, it *does* affect the clocks in the GPS systems that tell you how to get to the nearest florist. That's how nature works, and it is nature's gift—random as it may have been—that we evolved into beings with minds that can comprehend the difference.

DEEPAK

Eternity is in love with the productions of time.
—William Blake

Time gives spirituality a golden opportunity. People need a new way to live where time hasn't become a kind of psychological enemy. Deadlines press upon us. There are only so many hours in the day. No matter how fast we move, all of us are running out of time. Religion hasn't helped, because it tends to be grim about our time on Earth. What could be more depressing than the Puritan doctrine of "Sin in haste, repent at leisure"? If spirituality can free us from time's psychological downside, everyday life would be transformed.

Leonard is at pains to precisely define and measure time. He also hits upon a favorite point made by science, that subjectivity is unreliable. If you are a physicist collecting data on hadrons, bosons, and the like, you don't get to say, "My measurement changed because I have a migraine." But people don't use subjectivity to measure time; we use it to *experience* time. There is no other way. Time in all its aspects comes to us through our nervous system, as an experience in consciousness. Being conscious of time isn't abstract and objective. It's personal and participatory. And once we know how we participate in time, we will have an important clue about how to participate in the timeless.

The timeless? At this point I can imagine a wave of doubt coming from the reader, even a sympathetic reader. I'm not challenging the accuracy of the cesium-133 atomic clock, because there is no need to do so. Any aspect of time, including Einstein's relative time, is a by-product of the timeless; in the preuniverse there is no time. Our very source is the realm of timelessness. The story of how time sprang from eternity is the real mystery, and it's one that spirituality can solve. When you and I can experience the timeless, then phrases like "eternal life," "the immortal soul," and "a transcendent God" aren't just wishful thinking. When we look at it closely, eternity doesn't mean a long, long, long time. It means a reality where time is not present. But how do we actually get there?

Let's establish a point where spirituality and science agree. Time is relative. It isn't fixed. We don't need Einstein to confirm this, because everyday life already does. Depending on the state of consciousness you are in, the flow of time changes. In deep sleep there is no experience of time. In dreams, time is completely fluid: an epoch can pass in a moment, or a passing moment can last an epoch. (A story about the Buddha has it that he shut his eyes for a few moments, and yet inside he was experiencing thousands of years past.) Leonard stepped off this train before it left the station. He argues that time coming to us "through our senses" isn't the same as "the workings of the inanimate universe." But there is much more to consciousness than the five senses. Birds, bees, and snow leopards would see a mountain, the sky, or the moon in different ways because those creatures possess unique nervous systems.

If you change the nervous system, the idea of objectivity breaks down. This holds true not just for animals but for us, too. In recent experiments Buddhist monks were shown to have brain waves in the gamma region that were twice as fast as the norm: 80 cycles per second instead of 40 cycles. Gamma waves are thought to be the brain's way of holding the world together as a conscious experience. So Buddhist monks, by receiving twice the number of signals per second, are twice as awake, or conscious. Other people, operating on half-wakefulness, are sleepy and dull by comparison.

We can match this finding to other experiences. Quarterback Joe

Namath reported that when he was "in the zone," time seemed to stand still. The ball left his hand as if in slow motion, while at the same time the roar of the crowd disappeared, and he knew, with certainty, exactly where the ball would go; he even knew it would be caught. In other words, time cannot be detached from personal experience, which in turn means that no two people experience time in exactly the same way.

Subjective time, far from being an illusion, meshes quite well with post-Newtonian physics, where the notion of an objective observer was undercut long ago by relativity. If a space traveler's starship begins traveling near the speed of light, his time, as observed by someone standing back on Earth, slows down. This is a basic principle in relativity. Yet even as time became slower than molasses on a winter's day, as observed from Earth, the space traveler would register the clocks around him ticking off seconds, minutes, and hours in normal fashion. Likewise, since the gravitational field gets more and more powerful in the vicinity of a black hole, a faraway observer would see a space traveler's time seem to slow down until it virtually stopped altogether as the traveler approached the black hole's horizon—he would appear to take an infinite amount of time to cross that horizon and enter the hole. However, relativity is secondary to my main point, that some kind of nervous system is inescapable, and therefore so is the central role of experience. Science may not care, in objective terms, if Joe Namath felt time slow down; the timekeeper's watch says it didn't. It's up to me, then, to show how subjectivity is actually reliable. In India's spiritual tradition the zero state of consciousness is referred to as *samadhi,* where the mind enters pure consciousness. This state is an experience of an eternal timeless now. Here time does not exist as a measurable event. Only after pure consciousness splits into subject and object do we experience the flow of time.

Once again the findings of great sages mesh with quantum reality. (I apologize for giving the impression that all sages are either Indian or ancient. They cover the span of time, East and West. I give special weight to the ancients only because their spiritual observations have passed the test of time—whatever time turns out to be!) The underlying state of the universe is timeless. Before the first nanosecond of the Big Bang, there was only *the potential* for time in a dimension of all possibilities, after which quantum objects (e.g., energy, spin, charge, gravity) emerged. A potential doesn't have a life span. It encompasses past, present, and future. The ground state of physics turns out to resemble the zero state of *samadhi.* Once these timeless possibilities begin to collapse into space-time events, our connection to eternity seems lost. That is an illusion, though, fostered by our dependence on clock time. You have always been eternal; you still are.

There is certainly a huge objection to the claim that we can experience eternity. How can the human mind think about the timeless when thoughts take time? Everything human takes time, from being born to lying on one's deathbed. But the great sages noticed that the movement of thought is critical to time. If thoughts stop moving, so does time. We've all had a hint of this. When you say, "Sorry, I blanked out for a second," you weren't participating in time: the clock stopped. The Buddha took a more radical stance. He (and many other spiritual teachers) declared that when the mind stops, *everything* comes to a halt. It's not just time that is the movement of thought—the whole universe is the movement of thought.

Take this insight seriously and you wind up with an earthshaking idea: the state of precreation *thinks itself* into becoming the universe. Infinity transforms itself into the finite. Using whatever vocabulary you prefer, a silent mind (belonging to God, Brahman, nirvana, the absolute) creates physical reality through a thought, because without a vibration or frequency, time cannot begin. The same is true of space. Without some kind of vibration, there is no Big Bang, no expanding universe.

Vibrations emerge from a silent, motionless source. Then, as time enters creation, it is adapted to whatever nervous system uses it, including our own. Snails, for example, have a neural network that experiences time in wide intervals, as long as five seconds, as if a snail is seeing the world in a series of snapshots taken five seconds apart. If you reach down and snatch a lettuce leaf away from a snail fast enough, the hungry creature will experience the leaf vanishing into thin air. A snail can't make time speed up, but we humans have a special capacity: we can experience time at different speeds. Many versions of time are available to us, not just steady forward motion measured by the clock. We see the past repeat itself; we observe the cycle of life; we can take our imagination forward or backward; we feel time hang heavy, speed up, and even stop.

Medical doctors worry about "time sickness," a generic term for disorders resulting from the speed of modern life. Too much speed leads to stress, which in turn leads to the higher levels of stress hormones connected to many lifestyle disorders, like heart attacks and hypertension. Time literally runs out for a certain percentage of the newly widowed, or the chronically lonely, for whom time is such a burden that they run the risk of premature death. That's why it is so important not to merely describe time, as science does, but to understand it.

Changing your sense of self can give you more time, and improve its quality. A lot of research has been done on telomerase, a specific protein that seems to help cells live longer. The underlying theory is that telomerase keeps genes from unraveling and undergoing damaging mutations, so increased levels of telomerase may have a beneficial effect. Studies have shown that telomerase increases with positive lifestyle changes, and more than that, a person's sense of well-being—particularly the positive psychological changes brought about by meditation—promotes telomerase activity. (A coauthor of this study from 2010 was Dr. Elizabeth Blackburn, professor at the University of California, San Francisco, who shared the Nobel Prize in Medicine for the discovery of telomerase.) Just as we can alter the way we metabolize food, we have control over how we metabolize all experiences, even one as abstract as time.

It boils down to this: Human beings stand on the cusp between time and the timeless. We are a lamp at the door, to use an ancient Vedic image. At any moment we can look into the manifest or the unmanifest, the visible or the invisible, the world of time or the infinite expanse of eternity. Once we escape the mind-made trap science has unwittingly laid, we find ourselves granted enormous freedom and power, but this mastery over Nature is not an endorsement of the use of blunt force. Instead of coercing the physical world to do our bidding, we can use consciousness to achieve anything at all. Once our minds can travel back to the source, we recognize ourselves as part of the creative process that gives rise to space, time, and the physical universe. Here is the true power of now.

6

Is the Universe Alive?

DEEPAK

The possibility that we live in a universe that has a life of its own has intrigued human beings for centuries. Religion tells us that the universe is imbued with the divine force of the Creator; therefore it is alive. But I'm responsible for revisiting every concept with a mind that takes both science and spirituality seriously. This isn't easy, since science takes the position that the first primitive life-forms emerged 3.8 billion years ago, which is the same as saying that the Earth—and the universe—was dead before that moment. Why is it so necessary to make death the foundation of life, as if it is more real? That's what science insists upon.

Even more real than death, however, is flux. The cosmos is part of a never-ending process that recycles matter and energy. Nothing has a fixed identity: not a star, a galaxy, an electron, or a person—not you or me. Nothing, then, is truly dead. This isn't just philosophy but an observable truth. Every atom in your body came from an exploding supernova or interstellar gases; you and I are made of stardust. Our lives extend far beyond what happens to us personally, and at a subtler level, Nature is also recycling information and memory. Every time a cell divides, it must remember how to do that from the cells that came before it, which means that the molecules that produce enzymes and proteins inside a cell are programmed with the information, or code, for what to do.

You are the embodiment of a dynamic universe, which means you extend far beyond such narrow identities as "I am a Caucasian male" or "I am forty and happily married." Seeing yourself in any bounded way is illusory, just a wisp of thought drifting by in an eternal continuum. Spirituality provides a way to know yourself beyond the personal, which leads to enlightenment. I know this sounds lofty. To bring it down to earth, we need to build an argument on credible facts. The first fact is the one we've just discussed, that the universe is a living process, despite claims to the contrary.

Obviously we witness physical aspects of death all around us. But to equate that with death itself is shortsighted. Science and spirituality come to a decisive break in this regard, because science defines death in purely physical terms. Without a space suit, a human being (or any living thing, presumably) would die in the freezing vacuum of outer space within seconds. This fact, however, is irrelevant in determining whether the cosmos is animate. What's at stake in deciding between a dead universe and a living one is consciousness. If the cosmos is self-aware, as I've argued, it is alive.

Discovering consciousness in the universe is much more momentous than discovering gravity, although science doesn't seem to think so. There are good reasons for this resistance. In the materialistic scheme, matter must precede the emergence of life. The universe must be considered dead before DNA appeared. Even so, it seems like a miracle—or the remotest chance in the universe—that DNA learned to reproduce itself, a molecule that somehow reaches down and unzips itself into identical mirror images. No molecule had that ability before DNA appeared (although crystals are capable of simple replication, as children learn when they dip a string into a saturated sugar solution and watch sugar crystals begin to form on it, like stalactites in a cave). Spirituality doesn't need a miracle to explain life once the concept of a dead universe is discarded. What I want is to shed light, not make the case for magic. Far stronger is the argument that the universe gave rise to complex life because life has always existed, going back to the precreation state.

A cell that grows and multiplies looks much like a robot that has learned to build itself. Such a robot is logically impossible without a creator, since somebody or something had to assemble the first robot and program it. I apply the same logic to the cosmos. It creates itself, and if that is physically impossible without some kind of programming, then the miracle that DNA pulls off—self-replication—must be only one aspect of the cosmic program. At every second the universe disappears into the void and returns by re-creating itself. Physics explains this rebirth by pointing to the laws that govern the universe: they act like the meshed gears of a grandfather clock, only in this case the gears are invisible.

I am arguing that the recipe for life on Earth is wrapped up in the underlying existence of cosmic self-creation. The technical term used is "autopoiesis," literally "auto" (self) combined with the Greek word meaning "to make." No one can deny that the universe creates and maintains itself, just as a paramecium does as it floats on a pond in the sunlight.

On a cellular level each paramecium isn't a descendant of the first one that evolved billions of years ago: it *is* that first one. Completely identical versions are made by cell division, adding and subtracting nothing. It is true that new raw materials have to be collected to construct each generation of paramecia (and mutations may occur along the way, most dying out), but that is secondary. Life is like a house that keeps standing, looking the same from day to day, even though each brick is constantly being changed out for a new one. Food and air constantly fly in and out of every living cell, but *something* remains intact.

I can choose to call this invisible organizing power "life," but a specific explanation emerges only when we look more closely at autopoiesis, or self-creation. Four elements are involved, and I apologize in advance for how technical they sound. To be self-creating, you need:

1. A mechanism that is unified, with the ability to build itself
2. Component parts that self-organize into that mechanism
3. A network of processes that can transform themselves into anything the unified mechanism requires
4. A self-contained space that doesn't depend on an outside cause

This is much more abstract than saying "We reside in a living universe," although that is the conclusion to which these four requirements lead. Let me break them down by looking at an embryo gestating in the womb. The embryo is unified—we see one cell dividing into two, four, eight, sixteen, and so on, through fifty replications, all tending to the same goal: a baby. It grows as its components (food, air, and water) come together to serve the common goal. A network of processes constructs each cell, leading to another network that turns stem cells into specialized heart, liver, and brain cells. Finally, there is no need for an outside cause. The fertilized ovum can be put into a test tube, and even in such sterile isolation from the mother, as long as the first three components are provided for, a baby will begin to grow.

A skeptic will argue that the universe doesn't work this way. It only looks like a living organism. By analogy, the sugar crystals growing on a string when it is dipped into a saturated sugar solution aren't alive, even though they grow and reproduce. But autopoiesis can't be compared to crystals. The universe had no growing medium, no equivalent of a sugar solution. It created itself out of nothingness. Self-creation simply changes its costume when a baby is born. A baby, a galaxy, a photon, and the ecology of the rain forest look nothing alike, yet when you examine life at the deepest level, nothingness is creating every aspect of the living universe. Life is the universe's way of inventing eyes and ears to see and hear itself. The human brain is an observation deck for the cosmos to experience itself.

If you follow this path of inquiry, there is abundant evidence that the potential for complex life-forms has been embedded in the cosmos from the beginning. Since we will be discussing life at greater length in upcoming chapters, I will only offer a summary here in order to set the stage.

The universe can be understood as a living thing because of:

1. **Autopoiesis:** Any living thing grows from within.
2. **Wholeness:** Living things function as a single process, unifying many separate parts.

3. Consciousness: Living things, whether primitive or complex, exhibit awareness. Unlike inert chemicals, they respond to the environment.

4. Life cycle: Living things pass from birth to death, and in between they sustain themselves.

5. Spontaneous reproduction: Living things multiply and gather into populations. Within these populations there is a relationship among the individual members.

6. Creativity: Living things evolve; they don't mechanically reproduce clones. Thus we see a display of constant creativity.

7. Manifestation: An animate organism takes abstract ingredients and projects them into space-time, like a living hologram. These projections can be seen; they communicate; they enter the dance of life. When you break down any living thing, including the universe, you arrive at the abstract level again. Along the way, the spark of life seems to vanish. Examined under a microscope powerful enough to reveal molecular structure, living tissue can be reduced to inert chemicals. In reality, however, the spark of life didn't go out, because there was never a spark to lose. Life is in the void, too, but so abstractly that it takes a hologram—like you or me—to manifest it.

From a spiritual perspective, asking whether the universe is hospitable to life is a meaningless question. The universe and life are the same. We cannot be fooled by the mask of materialism. Behind the mask, the dancer is the dance, ever and always.

LEONARD

In 1944 psychologists Fritz Heider and Marianne Simmel made a short film featuring a circle, a large triangle, and a small triangle. The action involves these geometric figures chasing each other around until a final scene in which one moves offscreen and another breaks apart. You might think that such a movie would have the emotional resonance of a text on Euclidean geometry. But when Simmel and Heider asked subjects who had viewed the film to "write down what happened," the subjects made it sound like Academy Award material, interpreting the geometric figures as people, assigning the inanimate figures human motivations, and inventing a plot to explain the figures' movements. People like a good story enough to read one into just about anything. We anthropomorphize everything from cats and dogs to cars and, apparently, even our geometry, so it's easy to understand why a metaphysical theory about a universe that lives and thinks appeals to us.

Deepak offers an attractive story in which equating the physical aspect of death with the end of life is said to be "shortsighted," for we are all part of a universe that is "self-aware" and therefore "alive." In order for the statement that the universe is a living entity to have meaning, we must understand what it means for something to be alive. One can say that a piece of toast is alive, but try getting it to butter itself. A rock could be called alive, but chances are you'll never see one give birth. Usually when we think of something as alive, we mean at a minimum that it reacts to its environment and is capable of reproduction. What do these criteria mean when we speak of the universe?

Deepak lists seven requirements for life that he says the universe satisfies. First on the list is growth. Does the universe grow? To grow means to increase in size or substance. The universe is not increasing in substance, and physicists believe it is infinite, so the size issue is a subtle one. But if you place any boundary within the universe, that region will grow because, as I explained earlier, space is expanding. So the requirement of growth can be said to be satisfied. His second criterion, wholeness, requires that a living thing function as a unit. That's a squishy requirement. Pick your favorite sports team. Does it function as a unit? A good one does, a bad one doesn't, and coaches, writers, and fans can argue endlessly without coming to any definitive conclusion. But the universe by definition includes everything, so it would be hard to argue that the universe doesn't satisfy "wholeness." The life cycle requirement, that living things pass from birth to death, is satisfied by all objects that don't last for an eternity. The birth of a child is not the same as the birth of a chocolate cake, but still, one might say the universe also satisfies this criterion. On the other hand, most mainstream physicists would not say that the requirement of reproduction is satisfied. One might consider leaving this as an open question, since some untested and highly speculative models in cosmology—like the so-called ekpyrotic universe—come close to this, allowing that universes can be reborn, phoenixlike, from their own remains. But even in those models, the newborn universes don't "multiply and gather into populations" as Deepak requires, so one can only conclude the criterion of reproduction is not satisfied. The consciousness requirement—which is said to mean that an organism responds to its environment—cannot be applied to the universe, because the universe, being "everything," has no environment. Similarly—as I argued in chapter 4—because the universe does not exist in an external environment and undergo natural selection, it cannot be said to be evolving in the biological sense of the term. So the universe does not satisfy this criterion, either. Deepak's concept of a living universe is an interesting one, but these latter three criteria show that, even according to Deepak's own definition, the universe is not alive.

Could the universe be considered alive in some more abstract or generalized sense? Deepak talks about changes that happen within the universe, such as the development of galaxies and life, and asserts that "life is the universe's way of inventing eyes and ears." The real criterion for judging if the universe is alive, he offers, is not his checklist of the usual characteristics, but this: if the cosmos is self-aware, or conscious, it is alive.

Deepak believes that discovering consciousness in the universe is more momentous than discovering gravity, but "science doesn't seem to think so." Actually, science would think so. True, there might be the vociferous opposition that often accompanies new approaches. But if it were *discovered*—rather than merely proposed—that the universe is conscious, history shows that scientists would eventually swarm all over the discovery, and before long you'd have Nobel Prizes awarded and thousands of articles written on the psychology of the universe, with titles like *Are Supernovas Self-Destructive?* and *Are Black Holes a Sign of Depression?* Scientists make their careers on new and revolutionary ideas—especially young scientists, whose reputations do not depend on the continued usefulness of the old revolutionary ideas. But to gain acceptance in science the idea must have testable implications, and this concept of universal consciousness doesn't seem to.

The evidence Deepak offers is this: he says universal consciousness explains how life originated in the universe. We'll get to that claim soon, but first let's clarify the issue. Deepak compares the appearance of DNA to a zipper that somehow learns to reach down and unzip itself. Where did DNA come from? he asks. That is something that needs explaining. We know what happens once simple-celled organisms have formed: Evolution brings about the ever-developing progression of life-forms from simple to complex cells, then multicellular life, and then insectlike creatures, fish, amphibians and reptiles, birds and mammals, and finally primates, and us. But though evolution creates organisms of increasing complexity, all these organisms, going back to even the simplest bacterium, have something in common, which is that they are packed with molecular machines that create energy, transport nutrients, relay messages, build and repair cell structures, and perform many other amazing tasks. These molecules are mostly a type of molecule called an enzyme, which is a catalyst made from proteins (a catalyst is a molecule that changes the rate of a chemical reaction). Since all life utilizes such molecules, one might conclude that they are a requirement of life, at least of life as we know it. The issue is that if even the first simple living organisms from which everything today evolved included these structures, then how did these molecules first come into being?

The origin of life is an ongoing field of research, with many questions yet unanswered, but experiments suggest that it is possible for genetic molecules similar to DNA to form spontaneously, and other experiments suggest it is possible for those genetic molecules to curl up and act as catalysts. That means that the earliest forms of life, or what we might call "prelife," could have consisted of membranes made from fatty acids—another type of molecule known to form spontaneously—that enveloped a mix of water and those genetic molecules. Random mutations could then have taken over, enabling those cells to adapt to their environment, creating life as we know it today. Remember that even if the spontaneous origin of life, or prelife, within any given star system is improbable, that would not preclude its occurrence, because there are ten billion trillion stars in our observable universe. So as long as by "improbable" you don't mean less than a one-in-a-trillion shot, you could still expect over a billion star systems to harbor life.

Suppose life in any given star system *is* a trillion-to-one shot. How can we account for being so lucky? If, out of a group of a trillion stars, through the normal processes of nature, exactly one star system develops life, it might seem to the beings in that star system that their presence there is a miracle. Certainly if they chose their home by throwing a dart at a map of the heavens, the odds would be a trillion to one against hitting a life-bearing solar system. But that's not what happened. They were *born* into a star system in which life developed. And no matter how rare life is, by definition if living beings look around, they will find that they were born into a star system that harbors life. So that is *not* a miracle, or even good luck. It is just a consequence of logic.

Scientists may not yet have solved the problem of the origin of life, but our civilization is not so advanced in its discoveries that we should leap to the conclusion that if science hasn't yet been able to explain something, it never will. As the alternative to science, what does Deepak's metaphysics offer? How does a living, conscious universe explain how life appeared? He says, "Spirituality doesn't need a miracle to explain life once the concept of a dead universe is discarded. . . . Far stronger [than the appeal to a miracle] is the argument that the universe gave rise to complex life because life has *always* existed, going back to the precreation state." Such an argument might sound deep when applied to life and the universe, but let's examine the logic in a more mundane context—say, breakfast foods. Then the argument goes something like this: "We don't need a miracle to explain how the sunny-side-up egg appeared on my plate once the concept of an eggless plate is discarded. . . . Far stronger than the appeal to a miracle is the argument that the universe gave rise to a sunny-side-up egg because the sunny-side-up egg has always existed, going back to when the plate was originally manufactured." This explanation is obviously not very enlightening.

Deepak's argument is similar to that of Thomas Aquinas's thirteenth-century "first cause proof" of the existence of God, which goes something like this: Nothing can cause itself, so everything has a prior cause. Each prior cause must also have a prior cause. The only way to terminate this chain is for something extraordinary to exist which requires no cause, and that is God. God is that which can create, but which itself requires no creator. Even if one accepts that argument, it is a giant leap from this concept of God to Deepak's more specific concept of universal consciousness, or the biblical God that Aquinas employed this argument to justify. The argument really does nothing more than transfer the mystery of how a universe can come from nothing to the mystery of how God could have come from nothing. Simply asserting that God is God because God requires no cause does not get us very far.

After Stephen Hawking and I finished writing *The Grand Design,* I tried to describe the book to my then-nine-year-old daughter Olivia while waiting for a table at IHOP. Science tackles the big questions, I told her, and we want to explain the exciting answers to people who aren't scientists. *Where did we, and the universe, come from, and why is it the way it is?* She listened intently, and then I thought I'd check and see how much she'd absorbed. Why are we here? I asked her. She looked at me with an odd expression. Because we're hungry! she said. I guess I shouldn't try to discuss deep intellectual issues before breakfast.

We all have our own approach to the important questions, but once our hungers extend beyond a taste for pancakes to deeper human yearnings, we must be careful to start questioning the tooth fairy. The rigorous approach of science, which Deepak believes obscures the richness of life, is designed to help us avoid believing in seductive ideas that the evidence we reap from nature does not support.

Deepak writes that "higher consciousness allowed the great sages, saints, and seers to attain a kind of knowledge that science feels threatened by." We can probably all agree that the great sages, saints, and seers penetrated to knowledge that is outside the realm of science, and we can also agree that there are many kinds of subjective knowledge that are hugely important to us. It is important to know what makes one's children feel loved and secure and happy, for when, as an example, Olivia says that the adjective that best describes her is "joyful," this adds great meaning to my life. That such subjective experience is important does not threaten a scientist. But the danger of putting subjectivity on a pedestal and uncritically accepting metaphysical speculation as truth is that one will miss out on the most important intellectual understanding we can achieve—that of knowing the real place humanity holds in the physical cosmos. To me, that too is part of the richness of life.

PART THREE

LIFE

7

What Is Life?

LEONARD

Every spring in ancient Egypt the Nile River flooded neighboring land, and when it subsided it left behind nutrient-rich mud that enabled people to grow the crops that would sustain them. The muddy soil also gave rise to something else that wasn't around in drier times: a large number of frogs. The frogs came so suddenly that it seemed they had arisen from the mud itself—which was indeed how the Egyptians believed that they came into being. Medieval Europeans had analogous experiences. Butchers found that maggots and flies would soon appear on meat that had been left out in the open. Barnacle geese, which migrate by night, showed up suddenly on the coast of western Europe, apparently born from flotsam. Mice, too, seemed to generate themselves, in the grain stored in barns. In the seventeenth century a mystic and chemist, Jan Baptist van Helmont, even created a "recipe" for making mice: place dirty underwear in an open pot with a few grains of wheat, and wait twenty-one days. Though the theory was flawed, this was a successful recipe. For most of human history, it seemed obvious that simple living organisms could come into being spontaneously, a process that was called spontaneous generation.

But then different explanations began to emerge. In 1668 an Italian physician and naturalist named Francesco Redi suspected that the maggots that arose on meat—and the flies into which they developed—were due to tiny invisible eggs that other flies had lain. Redi performed one of biology's first truly scientific experiments to test his idea. He placed samples of snake meat, fish, and veal in wide-mouthed jars, leaving some uncovered, and covering others, some with paper, some with a gauzelike material. He hypothesized that if his theory was wrong, flies and maggots would appear on the meat in all three situations. But if he was correct, they would soon infest the uncovered meat, but not the meat covered by paper. He also expected to see flies buzzing around outside the gauze-covered jar, but not inside it. Later, he expected, maggots would appear on the gauze, and then drop onto the meat below. That is exactly what happened.

Redi's experiment threw a wet blanket on spontaneous generation, but the idea was not extinguished. With the development and refinement of the microscope, by 1700 people had for the first time been able to see all sorts of unfamiliar life-forms such as bacteria and other unicellular organisms. No one knew where these came from, but most people did suspect they were associated with the spoilage of meats and other foods that went bad. However, there were some who still favored the idea of spontaneous generation, because it seemed proof of a life force immanent in the universe. It could also be taken as evidence of how God could have created life from nothing. And so in 1745 a biologist and Roman Catholic priest named John Needham performed an experiment similar to Redi's, but on the microscopic scale. Knowing that heat killed the bacteria associated with spoilage, he heated chicken broth for a few minutes to kill anything that was living within it, then let it cool and sealed the vessel. A few days later, the broth showed signs of rotting. An Italian abbot named Lazzaro Spallanzani repeated Needham's experiments with a stricter protocol for sterilization, and the broth did not spoil. But Needham's experiment had breathed new life into the idea of spontaneous generation, and the abbot's more meticulous scientific work was not enough to kill it off.

The belief that there is some sort of essence—a life force—present in the universe was (and still is) appealing to many whose religious or spiritual views tell them that life is imbued with a special quality that can't be explained by the forces of nature. People had observed since ancient times that living things seem essentially different from the inanimate, so even apart from religious motives, it was natural to see in spontaneous generation evidence of some force that might be the carrier of this essence. About a century after the Needham/Spallanzani controversy, however, Louis Pasteur put the matter of spontaneous generation to rest through careful experiments that provided convincing evidence that microorganisms carried through the air, not born of the broth itself, are what causes broth to spoil.

So what is life? What does it mean to be alive? Deepak approaches consciousness as the foundation of a living universe. His views are reminiscent of a theory known as vitalism, which holds that life arises from a vital principle, or life force, that permeates the cosmos and lies outside the domain of chemistry and physics. If there were a life force that imbued each living organism, then the act of determining what is alive would be on the same footing as, say, that of determining whether an object is a magnet. Just as a magnet is a source of, and responds to, magnetic force, if there were a life force, a living object would interact with it and we could use that interaction to define and measure what is alive. But if there is no life force, then what is it that makes living things "essentially different"? How do we decide what is alive?

Biologists don't agree on the best way to define life. The living organisms we meet in our everyday world have some common properties, similar to the criteria Deepak gave in chapter 6: they undergo metabolism, which means they convert or use nutrients and energy; they reproduce; they grow; they respond to stimuli, such as when the leaves of a plant turn toward the sun; on a larger scale of time, their species change by adapting their characteristics to the demands of the environment; and they exhibit homeostasis, the self-regulating processes (having to do with everything from body temperature to the balance of biochemical substances in the bloodstream) that allow organisms to maintain a consistent internal state. For example, an ice cube tossed into a swimming pool is colder than the pool, but after a short time, it will melt and warm up, while the pool grows ever so slightly cooler. The forces of heat and cold, in other words, battle it out and come to equilibrium in the form of a uniform temperature. Similarly, a pot of boiling water placed in a cold stream will cool down, while the stream heats up ever so slightly, until the two reach the same temperature. A person tossed into a swimming pool or a cold stream, however, is capable of homeostasis and will maintain body temperature.

Though the above list of properties works well as a definition of life for turtles, redwood trees, and fungi, it is controversial in borderline cases like viruses, self-replicating proteins, and computer viruses. And who knows how exotic lifelike creatures we may someday discover on other planets might fit into our definitions? We've already seen that here on Earth, in an arsenic-rich environment, the sacred molecule of DNA operates in an alternate form, in which the phosphorus atoms in its backbone are replaced by arsenic, an element in the same family as phosphorus, yet quite different.

One can make a good argument that biologists don't need a single definition of life—the solution may be to accept that there are different categories of life, each exhibiting different combinations of lifelike characteristics. A virus may not satisfy all the traditional criteria, rock salt may satisfy just one or two, and a Martian microorganism three, but the details of how we choose to define life are unimportant as long as we are all aware of the criteria each of us is using.

Biologists want to know what makes living things tick, and so they need a definition of life for operational reasons. But here both Deepak and I are interested in a deeper question: what is the relation of living things to the physical universe? That is, if we consider squirrels, redwood trees, and fungi to be alive, and viruses, or even computer viruses, to be at least "lifelike," what physical qualities distinguish the atoms and molecules those things are made of, from the atoms and molecules in a chunk of metal, or sea salt?

If there were indeed a life force, one could say it instills into each of our molecules a quantum of vitality, making every atom within us alive. We'd be like a cake in which the sweetness of each crumb adds up to the sweetness of the whole. A living being, however, is not as alive as the sum of its parts. Life is what scientists call an "emergent property." An ocean wave depends on the interactions between many molecules, so to analyze a wave you must understand concepts like temperature and pressure that have no meaning when speaking of just a few molecules. Similarly, it is difficult or impossible, by studying individual molecules alone, to understand what it means to be alive. The atoms and molecules of something that has qualities fitting the definition of life are no different from those in a chunk of metal. It is only their organization that is different.

From the point of view of physics, living things are distinguished through their order, and their ability to maintain it. There are far more ways of rearranging the components in a pot of minestrone soup without destroying its identity as soup than there are ways to rearrange the parts of a cat without destroying its identity as a living thing, and so organization and order are more important to the cat than they are to the soup. Mess with how your molecules are put together, or which organs connect with which, and you won't last long. When we stop maintaining order, we die, and revert to a highly disordered state.

This idea was first popularized by Erwin Schrödinger, one of the founders of quantum theory, who gave a series of public lectures in Ireland that were published in 1944 as a book entitled *What Is Life?* I don't normally quote long-dead physicists, for a couple of reasons. For one, unlike religion, physics does not put much weight on authority. Certainly physicists listen carefully to the arguments of brilliant colleagues, but then we check their equations. More important, because science marches forward, every decent physics graduate student today knows far more than Schrödinger, Heisenberg, Bohr, Planck, Einstein, or any other pioneer of quantum theory ever knew about quantum theory, or any other fundamental theory in physics. And anyone who reads *Scientific American* knows more about the brain and neuroscience than they did. That doesn't mean that everything these scientists said was wrong; it just means that not everything they said was right, and for good and understandable reasons.

What Is Life? is famous in part because in it Schrödinger speculated about how genetic information might be encoded in living things. The book was later acknowledged as a source of inspiration by physicist turned molecular biologist Francis Crick, who with James Watson and Rosalind Franklin discovered the double-helix structure of DNA. In tackling the question posed by the book's title, Schrödinger also offered a pearl that still inspires the way physicists look at life, and describes that outlook very clearly:

> *What is the characteristic feature of life? When is a piece of matter said to be alive? When it goes on "doing something," moving, exchanging material with its environment, and so forth, and that for a much longer period than we would expect an inanimate piece of matter to "keep going" under similar circumstances. . . . It is by avoiding the rapid decay into the inert state of "equilibrium" that an organism appears so enigmatic.*

Living things are not like lifeless boulders rolling down a hill: thanks to homeostasis, our fluids keep their precise mix, our internal structures maintain their composition, and in warm-blooded animals, our temperature stays within a certain range.

When I talked about homeostasis I said a pot of boiling water tossed into a cold stream will cool down, while a human being won't. Of course, if you remain there for too long, your homeostatic mechanisms may be overwhelmed to the point that you develop hypothermia, and eventually die—at which time your body temperature will indeed be the same as that of the water, and you will be in equilibrium with your environment. However, most people will eventually feel uncomfortably cold and get out of the stream. So two of the fundamental characteristics of life are at work in thus resisting the fate of the pot of boiling water—metabolism (which helps you maintain your body temperature, at least for a while) and response to stimuli. That's life functioning on its most fundamental level—as a complex of energy-hungry molecules temporarily organized in a form that resists the inevitable return to equilibrium.

But the return is indeed inevitable. In this case I happen to believe rather literally what the Bible says in Genesis: "out of [the ground] you were taken; for dust you are and to dust you will return." Dust is a disorderly conglomeration of all sorts of tiny particles; but in between our beginnings from dust and our end as dust, the universe has given all of us living things the ability to maintain a strict order. For human beings this gift means that, for a time, our cells can stay organized and preserve the integrity of their content; our blood can flow through its proper channels within our bodies; our muscles, organs, and bones can maintain their structure and function. And, most important to our sense of who we are, it means that our brains can operate, and give us the capacity to reason, to store fond moments from childhood, to grow attached to others.

I spoke to my father while writing this book. For as long as I can remember I have feared for his health. When I spoke to him the other night he reassured me that he is alive and well, in the same way he has reassured me each time I've seen him over the last twenty years—in my dreams. My father died two decades ago but I'd obviously rather not accept it. I'd rather believe that he has rejoined the universe, or gone on living in some other form. Unfortunately, for me the desire is not strong enough to outweigh the skepticism. Deepak's metaphysics is not a religion, but like the answers of many religions, his answers are reassuring. It takes special courage to instead believe in science—to face the fact that after death our bodies return to the temperature of the inanimate objects around us, that we and our loved ones reach equilibrium with our environment, that we again become one with the dust.

DEEPAK

It takes a huge perspective to know what life is. If life arose from the most basic physical mechanisms that Leonard describes, such as homeostasis and heat exchange, blue-green algae would understand themselves better. But the rich depths of life haven't been plumbed by science, and that's what spirituality wants to address. In an earlier chapter, Leonard defended the superiority of science by saying that metaphysics can't build an MRI scanner. True, but the other edge of the sword is that metaphysics doesn't build high-tech weapons, either. Science can make life better in material ways, but no one could say that the world is suffering from a lack of materialism; in fact, the world is suffering from the exact opposite: a lack of self-knowledge.

Science could add to self-knowledge by expanding its sights. It could take heed of Einstein's core belief: "I maintain that the cosmic religious feeling is the strongest and noblest motive for scientific research." To my way of thinking, Einstein, Schrödinger, Pauli, and other so-called quantum mystics showed real wisdom in honoring the spiritual side of the human mind. After dedicating a lifetime to scientific research they came to the conclusion that spirituality offers a much broader exploration of life than science will ever come to on its own.

So, what is life? Life is the essence of existence. "Essence" doesn't mean a divine elixir that God poured into the ear of Adam and Eve. Nor is it the "life force" (more about that later). Essence refers to that which is most basic, the thing we cannot take away and still have creation. Evolution has given rise to millions of different forms, but let's not be distracted because plants and animals look different from stars and galaxies. Life is woven into the very fabric of the universe. You can't pet a star or walk an electron in the park, but deep down, they are both alive.

Why? Because, as we saw, the universe passes the same tests that biology applies to microbes, viruses, liver cells, white mice, and so forth. Every living creature is born and dies. The physical part decays and gets recycled into new life. Last year's fallen leaves become fertilizer for next spring's green buds. (It may make you queasy, but if a dead worm sends nitrogen into the earth, allowing an oak tree to grow, which drops acorns for pigs to gobble up, and you eat bacon for breakfast—well, draw your own conclusion about where your body comes from.) This cycle of rebirth isn't on automatic pilot, however. If an amoeba dies and decays, its raw materials don't have to come back as another amoeba. Any life-form, including the human body, can use those materials.

In other words, birth and rebirth are intensely creative. Something old and familiar leads to something new and original. The universe has been perfecting its creative abilities for billions of years. This creative drive is what I would call the "life force." Leonard maintains that real forces can be measured; some kind of meter, like the electrical meter fixed to the side of your house, must be able to measure it. But the life force is more like the power of imagination. If you measured the calories put out by Leonardo da Vinci's brain, you wouldn't be measuring the power of his imagination. His brain happens to give off heat, but that is a side effect, not the real power, which is invisible and immeasurable.

Materialists may tut-tut their disapproval, but forces exist that don't register on scientific instruments. (The force of desire, the force of curiosity, and the force of love could head the list.) Spirituality argues that creativity lies at the heart of everything that can be called alive. Does that mean that a rock in your shoe is alive? Yes, because it is part of the same creative process that includes you, a process that keeps endlessly emerging with new products. (It is fascinating to note that rocks needed life in order to evolve. The earliest phase of Earth's history began with 250 minerals, which came as we saw from the dust of supernovas and asteroid collisions. The turbulent forces on the Earth's crust, including the tremendous heat released by volcanoes, raised the number of minerals to around 1,500. But about two billion years ago living organisms began to process these minerals—feeding off them, and using them to build shells and skeletons. Tiny ocean plankton, whose skeletons are primarily made of calcium, laid down the White Cliffs of Dover and most other limestone formations. Amazingly, living things allowed minerals to keep evolving to reach the present number found on Earth, which is now 4,500—three times the original number. Cosmic evolution has relied upon life as a major cocreator.)

Leonard pleads with us not to fall for the delusions of metaphysics, as comforting as they may be: life is only the interval before dust returns to dust. But science has made a metaphysical decision of its own by putting its faith in matter. To say "We can do away with God" is metaphysics. To say "Life was created only by molecules" is also metaphysics. I'd call it bad metaphysics, actually. Basic physiology tells us that our brains are fed by glucose, or blood sugar. I couldn't write a word or have a thought without using up molecules of glucose. Yet even if a super-MRI in the future could match a molecule of blood sugar to the exact instant that a neuron fired the signal corresponding to a word on this page, it wouldn't mean that glucose is thinking.

Let's say that you track a brain cell back to the atoms that make it up, then farther back to subatomic particles, and finally across the divide into the invisible domain that lies beyond. No one can point to a specific physical process and say, "Aha, that's where thinking comes from" or "That's where glucose came alive." The effort to find such a starting point continues, but materialism is fooling itself. If a young child asked how gasoline learned to drive a car, he would be making the same mistake as some of our leading neuroscientists.

Every molecule that gets transformed into a living process poses an enigma. How does it go from an inert, random state (death) into a vital, creative state (life)? Spirituality takes the approach that nothing is dead. Because we fear our own disintegration and dissolution, we have projected onto death much more power than it actually has. Death is just a transitional stage, as one living form is reborn into another. (I'm not making a religious statement about the soul here, but I will later on.) Materialism can hypothetically trace the path of an oxygen atom in the jet stream until it enters the lungs of a future Michelangelo or Mozart, but it is useless in explaining how that atom is connected to genius, beauty, and art.

In order to explain how matter suddenly becomes part of the dance of life, with all the creativity that life exhibits, you must go to a more essential level. I've been arguing that consciousness is innate in Nature. It's part of our essence. So are the other qualities that distinguish life; intelligence, creativity, organization, and evolution are all essential to living beings. DNA didn't create them. Saying that DNA creates life is like saying that paint creates paintings. I believe we'll arrive at the truth by reversing the sequence: life came first, and eventually matter brought it into visible form. Physicist Freeman Dyson points the way to accepting the spiritual viewpoint as part of an expanded science: "I have found a universe growing without limit in richness and complexity, a universe of life surviving forever."

Some scientists seem to be willing to split the difference. Let biology tell us how life arose, they contend, while religion or metaphysics asks why. This is really a polite form of declaring victory, however, by giving life over entirely to science. Having identified DNA and having set out to map it, genetics is attempting to gobble up everything. There is supposedly a love gene, a criminal gene, even a faith gene. In fact no such genes have been found, and the leading speculation is that they never will be. An apparently simple problem like predicting a child's height involves more than twenty genes interacting with one another, and even if each of those genes could be isolated, researchers concede that less than half the story would have been told. Why have the Dutch shot up to be the tallest people in the world? Why are the Japanese now among the top ten? Their genes haven't changed. The answer lies somewhere with diet, environment, an unknown genetic switch, and perhaps an X factor (such as whether the mind can affect the body in its growth. Don't be incredulous—medicine already knows that psychological abuse can lead to stunted babies, through a process know as psychological dwarfism).

Science keeps getting greedier about the subjects it wants to gobble up. There is no room, as Leonard would have it, for wishful thinking, which we should have left behind in childhood. Don't talk to me about such fanciful things as intelligence existing everywhere. The best rebuttal I can offer is an eight-year-old border collie named Betsy who lives just outside Vienna, Austria. Betsy's owner trained her to fetch things by name. If she said "bone," Betsy fetched a bone. If she said "ball," Betsy fetched a ball. Any dog owner can tell you this isn't difficult, but this particular owner was more ambitious. She taught Betsy to fetch dolls, cheese, and a set of keys—until, against all odds, Betsy could understand 340 commands without getting them confused.

Cognitive psychologist Juliane Kaminski tested this phenomenon, which was filmed by public television's science journal, *Nova.* Human babies understand about three hundred words when they are around two years old. The next stage in human development, which no other primate has reached, is to grasp symbols. For example, if you hold up a tiny toy car and ask a toddler to find the same thing in the room, she knows that the tiny car is a model, so it poses no difficulty, at around age three, for a child to bring back a bigger toy car. News flash: so can Betsy the border collie. She understands that models represent things symbolically. (I can't resist mentioning that dogs are the only creatures besides human beings who know what pointing means. At six weeks old a puppy will go to an object if you point at it. At six months old, so will a human baby. But chimpanzees, our nearest primate relatives, cannot. If you point to a cup that hides a treat under it, chimps don't know what you mean. They don't catch on even after hundreds of repetitions.)

Betsy isn't the only smart border collie; at least two others can understand up to two hundred words, which runs counter to almost every old assumption about intelligence, the brain, the evolutionary ladder, and human pride in our exclusive mental abilities. Betsy has one more accomplishment to humble us with. It has long been claimed that only humans can understand abstract renderings. If I show you a picture of a bone, for example, you can run and bring me a real bone. So can Betsy. When shown a picture of any object she knows how to fetch, she goes and gets it. Researchers are left in awe, not before the grandeur of the universe but before an animal who has no right, scientifically, to do what she does. Yet she does it anyway.

Once we open our minds, Betsy can be the wedge to an all-embracing theory of life. The reader faces a clear choice between wholeness and parts. If science is right, life is a puzzle with lots of tiny pieces that, once assembled, turned inert matter into living creatures. If spirituality is right, life is part of Nature's wholeness, an aspect that becomes visible through living creatures but doesn't depend on them. The choice you make here reflects your worldview, and the universe will present itself accordingly.

The real problem with the theory of a life force is when it tries to be materialistic. But because it can't be measured, the "life" part of the life force has no material validity. Ironically, DNA runs into the same objection. I'm well aware that genetics is considered the greatest triumph in modern biology, the breakthrough that made it possible to decode life itself. DNA is the chemical carrier of an incredibly complex message, but it's not the message itself, any more than the letters of a telegram are the same as the thought that goes into the telegram. Life is Nature experiencing itself in as many different ways as possible. We can choose other words than "Nature." That's the message. We can speak of God looking at his (or her) creation, or the universal mind. Each term points toward a self-creating universe that unfolds as a living entity. Spirituality doesn't need a special moment when life suddenly appeared. Life has always been.

8

Is There Design in the Universe?

LEONARD

If by design one means a blueprint or pattern, then scientists and those with a religious or spiritual outlook can all agree that yes, the universe does have a design. We all see it with our eyes, and scientists seek to represent it through their equations, for we believe that the laws of physics are the blueprint for the universe. To create or simply understand a mathematical theory, and then observe as even the most minuscule atoms or the largest and most distant stars act according to the physical laws embodied in those equations, is one of the greatest wonders and joys of being a physicist.

Why nature follows laws is a mystery. Why the specific laws we've observed exist is also a mystery. But what is clear is that the laws of nature are sufficient to enable us to show how life arose without the necessity of there being any immortal hand or eye executing the design. Those laws dictated that from the primordial cosmic soup, stars would condense and create carbon and the other elements living things require. They dictated that some of those stars would then explode, and from the debris new solar systems would form. And they dictated that from the primordial chemical soup on at least one planet, ours, naturally occurring processes led to objects of beautiful design, from geodes to tigers to people.

The issue that separates Deepak and me is not whether the universe has design, but whether something designed it, and whether it was designed for a purpose. Creationists and adherents of "intelligent design" believe, as Deepak does, that the intricacies of living creatures could not be the result of natural law. That view has a long tradition. The British philosopher David Hume published a book in 1779 called *Dialogues Concerning Natural Religion,* in which three fictional characters debate the issue. One of them, Philo, puts the argument this way: "Throw several pieces of steel together, without shape or form; they will never arrange themselves so as to compose a watch."

In 1802 theologian William Paley famously elaborated on that theme:

> *In crossing a heath, suppose I pitched my foot against a stone, and were asked how the stone came to be there: I might possibly answer, that for any thing I know to the contrary, it had lain there forever: nor would it perhaps be very easy to show the absurdity of this answer. But suppose I had found a watch upon the ground, and it should be inquired how the watch happened to be in that place; I should hardly think of the answer which I had before given, that for any thing I knew, the watch might have always been there. . . . the inference, we think, is inevitable, that the watch must have had a maker; that there must have existed, at some time, and at some place or other, an artificer or artificers, who formed it for the purpose which we find it actually to answer; who comprehended its construction, and designed its use.*

The crux of all these gee-whiz arguments is that things as incredible as a watch or your grandmother are really complicated and hence could not have arisen except as the product of some being's exceptional expertise. They are sincere and compelling arguments based on the best science of their day, which was not up to the task of explaining how life came to be. But to paraphrase Arthur C. Clarke, any sufficiently advanced consequence of a scientific law that we do not yet understand is indistinguishable from the work of a "higher power."

Again and again through history people have assigned to any aspect of nature they could not explain an origin in the supernatural. Hume's character Philo was correct that pieces of steel thrown together will not form a watch, but that analogy seemed convincing only because people in Hume's day, nearly a century before Darwin published his great work, were not yet aware of the principle of natural selection, which makes it clear how unguided nature can indeed design amazingly complex objects (such as DNA; such as, ultimately, ourselves). Had a scientist from the future shown an eighteenth-century philosopher an airplane, an X-ray machine, or a cell phone, that philosopher would have been equally dumbfounded, and could just as well have attributed those devices to a divine origin. Perhaps then, some philosopher might have argued,

"Fix several wings together onto a steel hull; they can never be arranged so as to allow that hull to fly."

"Shine whatever light you wish onto a person's head; it will never allow you to peer at the brain inside."

or

"Scream as loudly as you wish into a tiny box; you will never be heard across the ocean."

Today science *does* explain how such devices can be constructed—just as it explains how natural processes lead to the development of intelligent life.

There is one difference between science's explanation of life and its explanation of those apparatuses. The science behind the airplane, the X-ray machine, and the cell phone doesn't threaten anyone's preferred beliefs. Hence, there is no public outcry against the science behind them. No one claims that scientists are closed-minded because they believe in aeronautics. No one proposes that X-ray images of broken bones don't really come from photons. No one says electromagnetism is "just a theory," or suggests that courses in telecommunications give equal time to carrier pigeons, just in case. But evolution concerns how we all got here, which makes it harder for some people to accept. The William Paleys of today willingly make use of the miraculous scientific feat that enables text messages offering two-for-one quesadillas to be coded into an invisible type of energy, transmitted through the air, and reconstituted on their handheld devices, but they question the integrity of the scientific method when it comes to the biological miracle of life. They are happy to employ inventions and products created through science they don't understand, but they balk at accepting the scientific "theories" that explain the very origins of life.

Biologists tell us that the designer of life was not a being, but the environment. The assumption implicit in the argument that complex things must have been designed by a higher intelligence is that it would have been simpler to accomplish the creation of life that way than through evolution. That is an understandable belief, especially for those who ignore the role of natural selection in evolution and view it merely as some sort of random hocus-pocus. But actually, because of the astonishing power of natural selection, the opposite may be true. That's why natural selection (technically, "artificial selection") has become the basis of a revolutionary new method of designing molecules called "directed evolution," in which chemists and chemical engineers set up environments that encourage starter molecules to evolve into commercially useful products. Directed evolution has proved successful in allowing the synthesis of many proteins that no one knew how to "design" in the traditional sense. So when admiring the amazing capabilities of life, perhaps it is more natural to say, not that they could only be the work of a creator, but rather that "this could only be the product of evolution."

Natural selection explains how organisms change from generation to generation until what started as the type of simple organism that causes stomachaches can evolve, after billions of years, into the type of complex organism that gets them. Darwin wrote about elephants. Suppose Noah had saved a single pair of elephants on his great ark, sometime around 3000 BCE, which was the time of the Flood. Though elephants are the slowest of breeders, in just five centuries they would have produced fifteen million descendants. By 2000 BCE, there'd have been trillions, many thousands of elephants for every person now alive. By now we should all have been crushed under a mountain of pachyderms. What saved us? Injury, sickness, starvation, and death. They ensured that only a fraction of elephants survived to produce offspring. It was not an unbiased pruning. On the contrary, in determining which should live and which should die, the environment acted as the intelligent designer. Animals that weren't tough or big or tall or smart enough to find sufficient food, fend off predators, and survive disease tended to die before they could pass on their ineffective traits. Those more suited to their environment survived and created progeny fit to compete in the next, new and improved, generation. And so on. In chapter 4, I mentioned that when a process like natural selection was included, in just forty-three generations evolution could create the Shakespearean phrase "Methinks it is like a weasel," which would take a random letter generator longer than the life of the solar system to produce. *That* is the power of evolution.

Evolution predicts that the design of living beings comes from both random mutation *and* selection due to the competition to survive. As a result, when studying living organisms in detail, one can't help but be struck by the fact that often their "design" is neither optimal nor elegant. It is, instead, "just good enough." Living organisms might be wondrous from the point of view of function, but they are not beautiful from the standpoint of design. That is very different from what you would expect if the design were created by an "intelligent designer," at least one that possessed superhuman intelligence. Evolution creates inelegant design because, as species evolve, nature doesn't tear down the house and rebuild from scratch, but takes the more expedient route of altering what's already there. Sometimes we're left with wisdom teeth or an appendix, or, as I'll talk about in the next chapter, a gene for a tail, traits that once served a function, but are no longer necessary. A purposeful designer would probably have made other choices, but since living organisms need not exhibit perfect design, evolution makes organisms that are just good enough to survive.

Evolution explains the origin of intelligent life on one level, but there is more to explain. Though biologists have made great strides in understanding the mechanism of evolution, right down to the molecular scale, biology is only the outer layer of the onion of scientific explanation. It describes organisms, their organs, cells, and, as of the last few decades, even the DNA, proteins, and other molecules living things are made of. But the descriptions and laws of biology take as their elementary elements objects which themselves can be broken down into more elementary components. At the deepest level—the core of the onion—lies physics. Physics is concerned with the forces and elementary particles that, by the trillions upon trillions, act to create the structures of the biologists' concern. So one ought to also ask, Does the development of life without the aid of a designer make sense on the level of physics? It is on that level that the answer to Deepak's challenge really lies: from the fundamental equations governing matter and energy, without any guidance or purpose, can life be spontaneously created? If we are to believe that no designer was needed, we must provide an answer that works not only at the level where biological processes are at work, but also at the level where the laws of physics operate.

To address whether, from the point of view of physics, the obvious design in nature required a designer, we must translate the issue into the language of physics. The early Earth was a rough mix of rocks and sand and air and water with various compounds dissolved or suspended within it. Living things, on the other hand, are made from very particular complex molecules and structures. The crux of the issue for physicists is: can such order arise without guidance? The tool physicists use to analyze that kind of question is a concept called entropy. Loosely speaking, entropy is a measure of the disorder in a system. The more disordered, usually, the higher the entropy. Entropy is the enemy of life, and of any concept of "design."

Physicists in the nineteenth century noticed that, with time, things tend to become more disordered—that is, the entropy increases. In a way this is a reflection of the lack of purpose or guidance in physical law. To understand why entropy, or disorder, increases, let's consider a simple (and classic) example, a box of gas molecules that has a partition down the middle with a hole in it. Suppose we start with a thousand molecules on the left side and none on the right. As the molecules bounce around, some of those on the left will pass through the hole in the partition and end up on the other side. With time, more will pass from left to right, but some on the right will occasionally pass to the left. That won't happen often as long as the right side is underpopulated, but eventually there will be many molecules on the right side, and so the net exodus will slow down. After more time there will be roughly the same number of molecules on both sides, and the number per unit time passing from right to left will be nearly the same as the number passing from left to right. That is an example of a state of equilibrium, as explained in the last chapter.

Though the term "disorder" is vague and subjective, it is probably safe to say that the initial configuration, with all the molecules congregating on the left, seems more ordered than the final one, in which the molecules are spread through the entire box. We think of the initial arrangement as ordered because it has a regularity—there are no molecules anywhere on the right side of the box. The final state of the box has no restrictions on its arrangement—the molecules are everywhere, so it is disordered. Our bodies, when we are alive, are like the initial arrangement. For example, our blood cells must maintain a certain internal biochemical balance, and not mix with their surroundings, and our blood must stay inside its vessels, and remain pure, not mixing randomly with other bodily fluids.

In the box scenario the initial configuration, with all the molecules on the left, was a low-entropy setup, and the final configuration, with the molecules all over the place, was a high-entropy situation. With time, and no higher consciousness or power at work to influence the distribution of molecules, the system moved toward a roughly equal split of the molecules, which is the most disordered, or maximum entropy state (that being the technical meaning of the term "equilibrium"). That is the tendency of all nature—the drive toward ever higher levels of entropy. As I explained earlier, life resists that drive. And when it ends, the drive toward entropy continues.

The law that explains why living things have to work at staying alive—i.e., at maintaining their order—is called the second law of thermodynamics. It dictates that the entropy of a closed system never decreases. That's a scientist's way of saying what Hume had his character say: "Throw several pieces of steel together, without shape or form; they will never arrange themselves so as to compose a watch." But the second law also says: "Leave a watch uncared for in nature and with time it will tend to become just several pieces of steel, without shape or form." The second law is why, if we shove a splattered egg off the counter, it will never hit the floor and coalesce into that nicely structured object we call an intact egg, but if we shove an intact egg off the counter, it *will* splatter into a random-looking mess. Similarly, if we find a box containing molecules equally distributed within it, we will never see the molecules all gather on one side, but if we find such a box with all its molecules on one side, with time they *will* eventually distribute themselves uniformly throughout the box. In view of this law, the challenge a physicist must address is, how can we start with atoms distributed willy-nilly throughout the universe, and find that at some later time they have coalesced into the ordered state we call living beings? In other words, if the natural tendency of the universe is disorder, then where does the order of life come from?

The phrase "closed system" here is the key. Entropy can't fall if there is no outside interference, but the entropy of one system can decrease if the entropy of another increases by an equal or greater amount. The hand of God may reach in and keep all the molecules on one side of the box, but that hand must suffer an increased disorder of its own. We keep the disorder in our bodies from increasing by consuming order in the guise of things like broccoli and chicken (until they've decomposed, they maintain a good bit of order) and expelling disorder as excrement and heat. So, too, must our planet respect the entropy balance. In order for life in our biosphere to have evolved from inorganic materials, the Earth needs to export entropy—that is, to import order. How? Where does the order come from?

Each day the Earth receives a sizable gift of energy from the sun, and also bequeaths a roughly equal amount of radiation back into space—that radiation balance keeps the planet's temperature from continuously rising. But the quality of the energy the Earth radiates is not the same as the quality of that which it receives. The surface of the sun is about twenty times the average temperature of the surface of the Earth, which means that the Earth must radiate twenty times as many photons—the particles of light—as the sun in order to radiate the same amount of energy. Physics tells us that this corresponds to twenty times the entropy, and so, day after day, the Earth radiates twenty times as much entropy as it receives. As Caltech physicist Sean Carroll has calculated, the net entropy generated by the Earth over the years is far more than enough to account for the entropy decrease the Earth has experienced by generating life.

The gift of life is not, then, the gift of a god, or of a "universal consciousness"; it is a gift from the sun.

DEEPAK

It's a shame that "design" became a buzzword for Christian fundamentalists, a pivot for their belief in the creation story of the book of Genesis. The word suddenly became radioactive in other circles. Scientists grew worried that reason itself was under attack. Skeptics and atheists threw their dogs into the fight, ever ready to beat back superstition. It thus became impossible to separate charged emotions from the issues that were at stake. Offering "intelligent design" as an alternative to Darwin's theory of evolution never did have any validity. What it did have was political clout. Elected officials who wanted to woo religious voters tried to sidestep overwhelming protest from the scientific community.

With this in mind, it's a welcome development when a respected scientist like Leonard agrees that the universe does, indeed, display traces of design. But his way of getting there is completely materialistic, meaning he relies on randomness and the dictates of the laws of Nature. There is a huge gap between "dictate" and "allow": without question the laws of Nature allow human beings to be here and to invent things like airplanes and watches, but did Bernoulli's principle, which allowed the Wright brothers to shape a wing in such a way that it got lift, dictate to them? The setup of the early universe cannot dictate my actions billions of years hence.

We take it for granted that there are ways to get around physical laws, usually by using one against the other. When I lift my arm I defy gravity by invoking electromagnetism, the force that controls muscles.

I can pull two magnets apart, using the same law against itself. As it exists today, the universe allows us enormous scope to play with the laws of Nature. Of course there are limits. I couldn't lift my arm on Jupiter, because my muscles would be too weak to counter that planet's stronger gravitational field. But materialism can't account for how a person chooses *which* laws to obey, counter, or play around with.

Latitude is built into Nature. When carbon, hydrogen, oxygen, and nitrogen meet up, their free electrons dictate that they will bond; all of life is based on such bonding, and as we observe, there are billions of possible combinations. Nature left lots of wiggle room for variation; therefore, the simple example Leonard provides of gas molecules drifting from the left side of a box to the right is not only reductionist, it doesn't apply. The same holds for the entire argument based on entropy. No one denies that entropy rules states of heat exchange. No one denies that living forms are islands of negative entropy. But the real mystery is how they got that way. The entire cosmos is heading toward heat death, as Leonard explains. But heat death is just a blown-up version of the molecules drifting in a box. Drift doesn't explain how islands of negative entropy, like the sun, the Earth, and life on Earth, can last for billions of years and keep growing more self-sustained.

Reductionism will always fail the test of how mindless natural laws can create anything as intricate as a watch. Leonard tries to escape the flaws in reductionism by verbal sleight of hand. He calls a watch complex, which it is. But it is more than that. It is designed. On the slopes of the Swiss Alps, one skier leaves a trail in the snow that is a simple line. A hundred skiers going down the same slope leave many more trails that form a tangled weave. The lines are more complex, but they are far from being a design. A Swiss watch doesn't just pile a bunch of simple processes on top of one another; it has purpose and meaning. It was designed to perform a specific task. It can be beautiful, but without a doubt it is precise. And when it drifts into inaccuracy, its imprecision can be corrected. All these aspects of design must have come from somewhere. Spirituality argues that they are aspects of consciousness, the invisible designer behind the scenes of the visible world.

I am not bothered when Leonard lumps my argument in with those of creationists and believers in "intelligent design." He isn't claiming that I am either of those things. Yet the lumping together does imply a kinship, which I must counter. Creationism and intelligent design are just as far from the world's wisdom traditions as materialism. When choosing sides in the ongoing debate between religious faith and scientific rationality, spirituality actually comes closer to science, since wisdom is the blossoming of reason, not its enemy.

I found it deplorable when a conservative White House announced that there was nothing wrong with teaching schoolchildren an alternative to evolution, and that children would benefit from an open debate. The public seemed to agree. In the end, it took the federal courts to affirm the obvious truth: intelligent design is a religious concept, not a scientific one, and therefore it cannot be considered an "alternative" in science classrooms. There is nothing to debate.

In an age of faith, the abundance of patterns in Nature was used to defend the existence of God. Leonard gives us the analogy of the watchmaker, which he associates with a kind of primitive, early scientific mind. That's not entirely right. The so-called argument from design was respectable on intellectual grounds in the seventeenth or eighteenth century. But it disappeared along with every other argument that tried to uphold the notion of purpose in the universe (known in philosophy as teleology). Scientists today offer the opposite, the argument against design, although they graciously allow that design can temporarily appear in the swirling randomness that rules all things.

The beautiful design found in Nature—as opposed to mere complexity and islands of heat—cannot simply be brushed aside. Science is forced to explain how design appeared in an accidental universe. For its part, spirituality is forced to explain the opposite, how randomness appeared in a purposeful universe. But if creation is imbued with consciousness, there is no war between chance and purpose, randomness and design. You can have both at the same time.

Look at your own life. You are a conscious being. Sometimes you stroll aimlessly looking at the scenery; sometimes you know where you are going. Sometimes you doodle and sometimes you draw. Aimless wandering doesn't negate destinations any more than a squiggle on a scratch pad negates studies in an art class. The same holds true on a cosmic scale. At a deeper level, random chance can benefit purpose. In the human sphere, letting go of a problem, releasing it to new possibilities, is often the best way to arrive at a solution. Nature seems to agree. The universe combines matter and energy, apparently by chance, only to arrive at sudden leaps of pattern and form. Before DNA there was a primordial soup of amino acids. The soup churned around without visible "design," but out of it emerged an incredibly complex design. This was creativity at work, not war.

Randomness can easily live in the same neighborhood as purpose, design, and meaning. All exist simultaneously in Nature. Red corpuscles bounce along randomly in my bloodstream, but I am not writing these words randomly. Being forced into an either/or choice—which is what happens when science says "choose materialism" and religion says "choose God"—puts a roadblock on the path to the truth. There is no use even arguing until everyone is willing to consider the deeper issues with an open mind.

9

What Makes Us Human?

DEEPAK

Darwin stands in the road as the enormous obstacle that religion never got around. So completely has the theory of evolution succeeded that most people cannot imagine a reasonable alternative. But it is possible to accept all the physical remains of our ancestors, tracing *Homo sapiens* back to early primates, and still derive a different answer for where human life comes from. Spirituality holds that the origins of human life lie in a transcendent realm beyond physical processes of any kind. We are mind first, matter later. According to Erwin Schrödinger, "What we observe as material bodies and forces are nothing but shapes and variations on the structure of space." If such a statement is true for the universe, it must be true for us, which means that space isn't empty; at the source, it's human (and a lot of other things, too). Jesus puts this much more poetically in the Gospel of Thomas when he says: "Split a piece of wood and I am there. Lift up a stone, and you will find me there."

So what does "human" mean, anyway? We are so complex and varied that there is room to view our species from any perspective you choose. I find it easy to sit in an armchair and agree with Hamlet when he exclaims,

> *What a piece of work is man, how noble in reason, how infinite in faculties; in form and moving how express and admirable, in action how like an angel, in apprehension how like a god!*

Suddenly I am transported back to the late Renaissance, to a world full of confidence, still holding on to the divine origin of human beings. But someone else could pick up an anthropology textbook and be transported just as quickly to the Afar Triangle in northeast Ethiopia, where paleontologists have dug up the oldest fossil remains of our hominid ancestors. Modern people are prone to believe in material things—skeletons, fossilized teeth, a hairline fracture showing where an animal attacked—as convincing scientific evidence. At the same time, bones and fossils discredited long-held concepts. It wasn't just religion that Darwin overthrew, but centuries of anthropocentrism, the belief that human beings are the most privileged creatures in creation. Suddenly we became nothing more than links in a biological chain. Lucy, the most famous example of *Australopithecus afarensis,* is a long way from Hamlet: about 3.2 million years. Every step back brings us closer to the animal kingdom and further from God's special dispensation.

But we'd be going to the opposite extreme to assess what it means to be human solely, or even mainly, from buried remains. It has been said that figuring out the human mind through physical evidence is like putting a stethoscope to the outside of the Houston Astrodome in order to learn the rules of baseball. Spirituality has no argument with paleontologists in their excitement over finding a hominid even older than Lucy (the latest candidate, announced in 2009, is Ardi, short for *Ardipithecus ramidus*—a male skeleton that dates to 4.4 million years ago, over a million years older than Lucy but far short of an undiscovered ancestor common to all hominids, targeted at around 10 million years ago). What spirituality does argue against is that any physical structure tells the whole story, either ancestral or present-day. Reductionism can follow the physical structure of the body down to the molecular and atomic level, but nowhere along that journey can physical traces tell us why we are creative, full of wishes and dreams, unique and different from one another, capable of memory, or many other things that are central to our story. Just as we need a Theory of

Everything for physics, we need a theory of everything when it comes to being human.

In asking where human life arose, spirituality has two advantages over science. The first, which sounds the simplest, is actually the most profound: spirituality embraces unpredictability. To the ancient Vedic sages, the whole universe was *Lila,* an expression of the playful, whimsical nature of God. The element of spontaneity cannot be dismissed from the human story. In the laboratory you can make mice happy by feeding them, and every time they take a nibble, a specific pleasure center in the brain will light up. You can go a step further and train mice to expect food every time they hear a bell or buzzer (a variation on the famous Pavlovian conditioning of dogs). When the mice simply hear this sound, the pleasure centers in their brain will light up, showing that mice can anticipate pleasure, just as we do when we think of an upcoming vacation in the Bahamas or the perfect Christmas gift. Brain structures in mice and humans are analogous, but this similarity proves very little, because on being shown a plate of food, a human being can say such things as "I'm on a diet," "It's too rare; I like my meat well done," "I'm too busy to eat," and "What about the starving children in Africa?" We humans possess countless responses to the same stimuli. No model of the human brain can predict which response you or I will choose, not just to food, but to anything at all. Unpredictability destroys all forms of determinism, and that's fatal for physical explanations, because physical systems are ruled by fixed processes. A carbon atom can't choose whether or not to bond with an oxygen atom. When they meet, their interaction is determined. When two human beings meet, they might not have any chemistry at all!

If you ask where unpredictability entered the evolutionary record (e.g., who was the first human to say, "You can have my mastodon rib, I'm not hungry"?), scientific answers do come back. We hear about selfish genes and altruistic genes causing us to behave in certain very human ways. But even if you could pinpoint a gene for selfishness and the opposite one for altruism, wouldn't we need a third gene to choose between them? After all, we can be both selfish and selfless.

Where is the gene that shows me how to select this word from the thirty thousand–plus in my vocabulary, or the chemical reaction that dictates where I will eat lunch when faced with a choice of a hundred restaurants in a midsized city?

The second advantage that spirituality enjoys over science is that it embraces the richness of experience. You can break any brain response down to action and reaction, stimulus and response. Imagine a lemon with a knife sitting next to it. In your mind's eye, see a hand pick up the knife and cut the lemon in half, then watch the juice being squeezed out. Almost everybody will salivate while doing this exercise, which to a reductionist means that we are like Pavlov's dogs, who salivated when they heard a bell. But dogs don't salivate to *imaginary* lemons, while we do that and much more: we create whole worlds in our imagination. The richness of inner experience encompasses everything human; it also defines us. We thrive on meaning, and we languish and atrophy in its absence.

Neuroscience probes for these qualities in brain tissue. Its worldview, and its methods, require such an approach. This gives rise to a strange blindness. Reductionists cannot, in my experience, be talked out of believing in a world where physical processes can eventually explain meaning, purpose, and all the rest. They would be better served by realizing a simple fact: you can't start from a meaningless cosmos and get to the rich meaning of human life. Spirituality turns the telescope around and looks at experience first. Then, if you ask where human life originated, your answer is that what really matters has no beginning or end. Human life is embedded in the domain beyond space-time, like everything else. From the Gospel of Thomas comes this passage: "If they say to you, 'Where have you come from?' say to them, 'We came from the light, the place where the light came into being of its own accord.'" The beauty of this passage is that it is equally true for science and for spirituality.

LEONARD

In 1522 townspeople in the district of Autun, France, were angered to discover that rats had eaten their crop of barley. The rats did not own the barley, and had not been authorized to eat it. The villagers went to court and obtained a summons ordering the rats to stand trial. Sounds strange, but in Exodus it says, "If an ox gore a man or a woman that they die, then the ox shall surely be stoned," so why should rats be above the law? Indeed, according to records, throughout Europe from the ninth through the nineteenth century, a wide variety of animals that violated human laws were put on trial just as people were. Oxen, pigs, and bulls were jailed, tortured for confessions, and even hung by the same hangmen who executed humans. In Autun a court official went to an area of the countryside where the alleged offenders were believed to reside, and read, quite loudly, a solemn notice demanding that the rats appear in court. When they didn't show up, a court-appointed attorney for their defense argued that more time was needed for them to make the journey to the courthouse. When they again failed to appear, their attorney argued that they could not be expected to risk death by hostile cats in order to obey the summons. These trials were not really about revenge against evil animals. Legal systems are about more than punishment and deterrence. They are about maintaining the social order, and in these cases the need to follow society's rules trumped any doubts that birds have souls, bees are capable of evil intent, or field mice can engineer fraud.

Our organization into social networks is a distinguishing feature of our species. Certainly social order is found not just among humans, but also among animals like ants, termites, and bees. One of our fellow mammals also lives in highly organized societies—the naked mole rat. Naked mole rats make their homes in subterranean hives supported by a specialized workforce, and sustained by a single breeding queen. On its own, a naked mole rat could not keep warm, obtain food, or avoid predators, and so it wouldn't last long. But even the well-socialized mole rat, upon bumping into others of its species, does not wonder if their search for food has left them stressed out, contemplate how they feel about the predator situation, or raise the issue of the starving rodents in Africa. A human being, on the other hand, might help an elderly stranger across the street, wonder how another person is feeling, or not trust a doctor who wears a nose ring. And humans have developed culture, which other species have only in rudimentary form. People are natural mimics, so even when we still lived in the wild, we were capable of learning new things, actions that went beyond the instinctive, by watching one another, an advantage most other species don't enjoy. It might have taken bears thousands of generations to evolve their thick fur, but all our species needed was a single human to get the idea that we could skin a bear to make a fur coat, thus enabling our species to stay warm forever after. Today we build on human discoveries made over the course of millennia, and share our knowledge worldwide.

The bonds cementing human society are far more complex than they are among other animals. Even among our closest mammalian relatives, our social abilities stand out. The taxonomic family to which human beings belong is called the hominids, and our genus, a kind of "subfamily" of closer relatives, is called *Homo.* Our species, *Homo sapiens,* is one of more than a dozen within the *Homo* genus, the best-known of these besides ourselves being Neanderthals, *Homo habilis,* and *Homo erectus,* all of which, of course, have long since died out—possibly because of the lack of those social abilities. Many of these nonhuman species engaged in humanlike activities, such as using tools, harnessing fire, interring the dead, and engaging in cultural rituals such as painting their bodies. But none lived in societies nearly as complex as our own.

What unique talents have we humans developed to enable us to interact so effectively with so many others, to live in cities exceeding a million or even ten (or more) million inhabitants? One is language. Not only does language facilitate intensely social interactions; it also allows the transmission of knowledge across society, and through the generations. Dolphins and monkeys might exchange signals, but only humans have the capacity to explain complex nuances to our children. A moral code is also important. Our primate ancestors may not have had to worry about a society gone amok due to investment fraud, but in general people living together have always done better if they have felt a reluctance to bash one another's skulls with a rock. It might seem we humans are always at war, but our hesitance to kill is actually so strong that a U.S. Army survey during World War II concluded that 80 percent of field combatants couldn't bring themselves to shoot at the enemy, even when attacked.

Human beings also engage in altruism far more deliberately and pervasively than any other species, and certain structures in our brain linked to reward processing are engaged when we participate in acts of mutual cooperation. Even six-month-olds evaluate others based on their social behavior. Infants in one experiment observed a "climber" consisting of a disk of wood with large eyes glued onto one of its circular faces. The climber started off on a hill, and repeatedly tried but failed to make its way to the top. Occasionally, after some time passed, a "helper triangle"—a triangle with similar eyes glued on—approached from below and aided the climber with an upward push. Other times a "hinderer square" approached from uphill and shoved the circular climber back down. The experimenters were investigating whether infants, unaffected and uninvolved bystanders, would cop an attitude toward the hinderer squares. And, judging by the infants' tendency to reach for the helper triangles rather than the hinderer squares, they did. Moreover, when the experiment was repeated with either a helper and a neutral bystander block or a hinderer and a neutral block, the infants preferred the friendly triangles to the neutral block, and the neutral block to the nasty squares. Long before we can verbalize attraction or revulsion, we have a sense of social morality—we are attracted to those who are kind and repulsed by the unkind.

Another quality that distinguishes humans from other species is our desire and ability to understand what others of our species think and feel. That ability is called "theory of mind," or "ToM" for short. ToM allows us to make sense of other people's past behavior, and to predict how their behavior will unfold given their present or future circumstances. Only humans have social organization and relationships that make high demands on an individual's ToM, and though scientists are still debating whether nonhuman primates use ToM, if they do, it seems to be at a rudimentary level. In humans, though, simple ToM develops in the first year, and by age four, nearly all children develop the ability to assess other people's mental states. It's what enables us to form large and sophisticated social systems, from farming communities to large corporations. When it breaks down, as in cases of autism, people can have a difficult time functioning in society.

All these qualities—especially ToM—require a certain amount of brain power, and so the survival advantages of social interaction may be an even more important factor in the evolution of the human brain than the skills and decision-making abilities the brain makes possible.

The capacities we've been discussing go to the heart of what makes us human, and we are getting ever more sophisticated in our ability to map the areas of the brain that are responsible for them. But Deepak looks to something less tangible for the source of our humanity, something that goes beyond the physical.

Deepak argues that spirituality has the advantage of including unpredictability and spontaneity as key elements in the "human story." He says that to look for the physical basis of humanity's essence will fail, because we are unpredictable, and "unpredictability destroys all forms of determinism" and so is "fatal for physical explanations." That's not in fact true. Quantum theory, for example, is famous for the limits it places on predictability, and physicists do fine with that.

Even without resorting to the esoteric laws of quantum theory we can find many examples of unpredictability that do not violate the laws of the material world. One such example is the dwarf planet Pluto, which has been shown to follow a chaotic orbit, meaning that its path, in the long run, cannot be predicted—but that doesn't mean Pluto is disobeying Newton's laws. Or consider the path of a simple stone rolling down a rocky hill. No physicist believes that he or she can predict it, but neither would anyone believe the path taken by the stone is beyond physical explanation. A hurricane, in taking its unpredictable path, *seems* to have a mind of its own, but it doesn't really.

The real issue in Deepak's argument is free will, and though the question of free will has important implications for our view of ourselves, from the practical standpoint, it is actually of questionable relevance. That's because whether or not people have free will in principle, in practice it appears that we do because our behavior is so difficult to predict. There is no contradiction in saying that our decisions are determined by the laws of physics, yet we still don't know how to reliably predict behavior. Human beings, like the dwarf planet Pluto, may well be so complex that our actions and decisions will forever remain to some extent unpredictable. But to say we cannot predict people's actions is a statement about our powers of prediction, not a statement about whether we have free will.

Deepak writes that a carbon atom can't choose to bond with another carbon atom, but (he implies) what makes humans special is that we *can* choose, that we have free will. Free will is a tremendously fraught issue. Modern psychology and neuroscience have addressed it employing a range of techniques, from direct electrical stimulation of the brain to cutting-edge neuroimaging and animal neurophysiology. And indeed, science is challenging our intuitive and traditional understanding of human choice: experiment after experiment seems to indicate that our choices are far more automatic and constrained than we'd like to think. Take your taste in facial beauty. That feels very personal, defined by your individual sensibilities, though perhaps influenced also by the culture in which you live. But numerous studies show that men and women, regardless of culture and independent of race, generally agree on which faces are most attractive—and that these preferences emerge very early in life. The key? Faces which possess features closest to average are considered the most appealing. So if you want movie star material the recipe is simple: toss one hundred random male or female faces into a specialized computer graphics program, and average them. It's not romantic, but it works—the faces that result from such manipulations are those we find beautiful. Our sense of morality, too, seems to be largely hardwired. Studies show that when confronted with a situation where questions of morality might arise, people seem to very quickly, and unconsciously, reach a moral judgment first, and only a split second later consciously construct reasons—based either on practical or on religious values—to justify how they feel.

The evidence so far supports the view that the physical arrangements of all the atoms and molecules, and the laws of nature that govern them, determine our future actions in the same way that they determine the actions of the sun, or the growth of a rosebud. But science has not *proved* that there is no immaterial consciousness that makes our decisions, nor is it clear that it ever can prove the absence of a phenomenon, such as a "soul," that has no physical manifestation. All science can really say is that if it existed, we think its effects on the material realm would have been noticed, and that, until now, there has never been any credible evidence for it.

It can be difficult to accept that nature, rather than some version of an immaterial self that transcends nature's laws, governs our actions. And it's very hard to see ourselves accurately, objectively. Our judgments are all made in the framework of our prior beliefs and expectations, which are themselves often influenced by our desires. Illusions expert Al Seckel offered me a striking demonstration of how expectation can shape beliefs. It started out with an excerpt from a Led Zeppelin song:

If there's a bustle in your hedgerow, don't be alarmed now,
It's just a spring clean for the May queen.

The next few lines go on to say that though there are various ways to live your life, you can always change your direction. After playing the song Seckel played it again backward, an effect easily achieved employing sound-editing software. It seems absurd to expect that a singer's voice could make linguistic sense played *both* forward and backward, and indeed, I listened to the backward version several times, and just as I assumed, it sounded like total gibberish. Seckel, however, claimed that this song *did* make sense when played backward, and that Led Zeppelin intended it to. To help me hear the message he said was encoded in that version, he offered me a framework—a printed-out version of the text to the backward lyrics, so that I could read along as I listened. Here is what it said:

Oh here's to my sweet Satan. The one whose
Little path would make me sad, whose power is Satan. He'll
give those with him 666, there was a little tool shed where
he made us suffer, sad Satan.

I expected that when listening yet one more time, I'd find that the backward song still sounded like gibberish, but this time, when I followed the text, it was striking how closely the words really matched. I was now convinced that Seckel was right, and I had a difficult time understanding how I could have missed comprehending those words the first few times! I was astonished. Then Seckel told me Led Zeppelin didn't really encode a satanic message. Seckel said they were made up. One could have made up other sets of words that fit the gibberish, he said, and I would have believed the song said them, had he provided those lyrics to me instead.

When perceiving the world without prejudice, as I had at first, our minds judge the world quite differently than when assessing it in the context of a belief or expectation, as I did after Seckel gave me the text. That is also true of the way we perceive ourselves. Our "self" is the most fundamental element of our world, and we do not approach the subject of "me" without bias or prejudice. Is our intuitive feeling about the special place our species holds in the universe (and about the free will that makes us so special) correct, like our understanding of the song lyrics? Or is it an illusion of our subjectivity, like our understanding of those lyrics when played backward?

How would we judge ourselves and humanity from the outside, if we weren't one of them? Advanced aliens would probably group us with squirrels and mice—lower beings that are mere automata—while perhaps believing that they themselves are different, that they are the only truly intelligent species, and the only species with free will. But according to the evidence of science so far, they too would be wrong. We are all governed by the same physics, the physics of this material world. I admit it feels strange to think of myself as a biological machine governed by the same laws that govern Pluto. But understanding my essence doesn't diminish my appreciation for the gift of being alive; it makes me appreciate it even more. That's not a scientific principle. It's just the way I feel.

10

How Do Genes Work?

LEONARD

On April 25, 1953, two young researchers at Cambridge University in England—James Watson and Francis Crick—published a paper in the journal *Nature* arguing that the structure of DNA consisted of two interlocking strands arranged in a double helix, something like a twisted chain ladder. In their proposed model each rung of the ladder consisted of a molecule called a base from one chain, paired with a complementary base from the other chain. As a result, if you pulled the chains apart, each of them could act as a template from which a new complementary partner could be created. In this manner, one molecule of DNA could turn into two. Watson and Crick's article was brief and contained only one sentence that hinted at its implications: "It has not escaped our notice that the specific pairing we have postulated suggests a possible copying mechanism for the genetic material."

Watson and Crick's publication came almost exactly two years before Einstein's death. Unlike Einstein's general relativity, their work was neither a great conceptual leap nor an advance that would have been greatly delayed had they not gotten there first. But it did mark the beginning of a new era in biology, which allowed scientists to study the details of inheritance on the molecular level. No one knew where that investigation would lead, although Watson and Crick published a speculative paper about the meaning of their work a month later. In June the *New York Times* ran an article with the timid headline "Clue to Chemistry of Heredity Found," along with a cautionary statement from famed Caltech chemist Linus Pauling that he "did not believe the problem of understanding molecular genetics had been finally solved." Pauling—who the next year would win the first of his two Nobel Prizes—was right.

How complex is the mechanism of heredity? Today, some sixty years later, tremendous progress has been made, but thousands of scientists are still working out the details.

The idea of evolution goes back at least to the ancient Greeks, but what many consider the first coherent theory of the subject—involving the concept of inherited traits—was proposed around 1800, decades before Darwin, by the French scientist Jean-Baptiste Lamarck. According to Darwinian evolution, new traits, such as the giraffe's long neck, arise through mutations, which make it possible that the traits of a child might not correspond to the traits of either parent. If, given the environment, that new trait turns out to provide an advantage, the child will thrive, reproduce, and pass the mutation on to subsequent generations. But Lamarck believed that animals' traits are not limited to the effects of their heredity. He proposed that traits can change during an organism's lifetime in order to allow it to best adapt to its environment, and that the organism's newly developed traits can then be passed to the next generation. In this view, for example, if a giraffe were suddenly moved to an environment with taller trees, its neck might grow longer, in which case subsequent offspring could be born with longer necks. Today we call that process soft inheritance. It is not the way evolution normally operates, though recently scientists have discovered that such processes do occur, spawning a field called epigenetics, which I will return to later.

Both Darwinian and Lamarckian theories of evolution raise a crucial question: how are traits passed from parent to child? In 1865 Czech monk Gregor Mendel presented a paper showing that certain traits in peas, such as shape and color, are passed along in discrete packages we now call genes, but his work went unappreciated until the turn of the century. Meanwhile, the molecule we now call DNA was discovered in 1869 by Friedrich Miescher, a Swiss physician studying white blood cells he obtained from the pus in surgical bandages. Miescher didn't know what the substance was good for, but he knew there was a lot of it—there is in fact enough DNA in almost every human cell to make a strand about six feet long.

The connection between genes and DNA didn't get made until 1944. Before that, if there was one thing scientists were confident about, it was that DNA was *not* the molecule of heredity. That is because DNA seemed far too simple—it was known to be made of just four different components, called nucleotides. (Each nucleotide consists of a base, as I mentioned—one of four different types—plus two other small molecules, a sugar and a phosphate molecule, which we now know form the spine of the DNA.) Then in 1944, after many years of intricate experiments, a shy sixty-seven-year-old researcher named Oswald Avery and his colleagues showed that if DNA was extracted from dead bacteria and injected into a live strain, it caused permanent changes in the DNA and traits of the live strain, which were inherited by subsequent generations. Avery's work inspired a search to discover the structure of that mystery molecule, culminating in Watson and Crick's discovery of the double helix in 1953.

Roughly speaking, in modern parlance a gene is a region of an organism's DNA that contains instructions for making a particular protein. Biologists say that the gene "codes for" the protein. The code, or recipe, is written using just four letters—A, C, G, and T, which stand for the four bases that make up DNA—but the recipe book is a long one, containing over three billion pairs of bases. When the recipe is used successfully to create the protein product, the gene is said to have been "expressed." The proteins are all cooked up from a pantry of just twenty amino acids. Proteins constitute much of any organism's physical structure, are involved in virtually every cell function, and control all the chemical processes inside the cell. Each of our bodies contains over a hundred thousand different proteins, including our hormones, enzymes, antibodies, and transport molecules such as hemoglobin.

The traits we inherit are determined by the proteins our bodies produce, which are in turn dictated by the recipes in our genes. The cookbook containing all those recipes is an opus of many volumes called the genome, the different volumes of which are called chromosomes. We all have distinct characteristics, some due to our environment and experiences, others arising from our heredity. Since each of us has a different heredity, my genome is different from yours. What, then, does it mean to speak of "the human genome"?

Our personal differences seem great to us. Some of us would rather shovel snow than listen to opera, while others can't imagine a world without *La traviata.* Some propose marriage over a quiet picnic on the beach, others at a table next to a drunken rugby team at the Outback Steakhouse. But on the level of genes, what makes us alike is far, far greater than what makes us different: the genomes of any two human beings typically differ by only about one letter out of each thousand. They are virtually identical, like copies of the same book that differ only in their misprints.

The misprint metaphor is apt here: our genetic differences arose through mutations—random changes in the genetic letters—that occurred over the millennia. These mutations account for that part of human variability that is not due to differences in experience or environment, such as our differing blood types, hair and skin colors, and facial features, and perhaps even for why some of us can carry a tune, while the singing of others could be used to keep rats out of the basement.

All told, humans are now thought to have about twenty-three thousand genes. That's fewer than a newt has, or a grape, which is bound to make those who believe that bigger is better a bit uncomfortable. That illustrates the dangers of oversimplified thinking, and indeed though I've given the big picture of how genes are connected to traits, it is important to keep in mind that it *is* a greatly simplified version. For example, each cell has not one, but two copies of the recipe book, since we receive one intact genome from each of our parents. When the recipes conflict, sometimes one prevails over the other, but at other times some sort of compromise is made, or a completely different protein is created. Also, many genes contribute recipes for more than one protein—almost half of our genes are spliced in order to produce multiple proteins, which is why we can have more than a hundred thousand proteins but just twenty-three thousand genes.

The effect of a gene also depends a great deal on what is called "gene regulation"—processes that determine whether the recipe dictated by the gene is actually carried out or, as we say, expressed. On the molecular level gene regulation occurs when certain chemicals interact with parts of the DNA molecule to inactivate a gene. As a result, for example, two identical twins—who by definition have the same DNA—can be strikingly different. In rodents called agouti mice one twin can be thin and brown, while the other is obese and yellow. Such obese yellow mice result from environmental effects. These mice occur occasionally in natural conditions, but when pregnant agouti mice are exposed to a chemical called bisphenol A, present in many plastic drink bottles, significantly more obese yellow mice are born. It was found that as a result of the exposure, the DNA of the offspring have less "methylation," a process that turns off genes. This causes more than the usual amount of a certain protein to be produced, which in some mice has two disparate effects—one in the skin (blocking cells from making black pigment) and the other in the brain (affecting feeding behavior). Though giraffes don't develop long necks by stretching toward trees, as Lamarck believed, the expression of genes—and hence the makeup of an individual—can, through gene regulation, be profoundly affected by the environment, and you don't need chemical toxins to do it. Himalayan rabbits, for example, carry a gene required for the development of pigment. But the gene is inactive in temperatures above 95 degrees Fahrenheit, which is below the rabbits' body temperature—except at their extremities, which are cooler. As a result, Himalayan rabbits are white, with black ears, nose tips, and feet.

Changes in traits that, like these, are due to mechanisms other than a change in the underlying DNA are called epigenetic. Because of gene regulation and epigenetic changes, there can be many characteristics within an organism (of any species) which did not arise at conception, but rather are a reflection of the interaction between the genome and the information in the organism's environment, from its time in the womb onward through life. In a few cases these epigenetic changes have been observed to continue through many generations. These instances correspond to the Lamarckian view of evolution, in which traits that change within the lifetime of an individual can be passed on to that individual's descendants.

Another complication to the simple picture is that only 1 or 2 percent of the genome corresponds to the genes I described above, the recipes for proteins. The rest was mislabeled "junk DNA" by scientists before anyone understood its purpose, but it has since been discovered that most of this "intergenic" or "noncoding" DNA—terms now preferred by scientists—does indeed serve an important function. About half of it stabilizes the structure of the chromosome, which is a strand of DNA packaged in a protein. Other sequences define where genes begin and end, something like the capital letter and period that play the same role in language. Sequences called pseudogenes are copies of normal genes that contain a defect that prevents their expression as proteins. They used to be thought of as vestigial—perhaps the only true "junk" in our genome—but a breakthrough in 2010 indicated that they may play an important epigenetic role, keeping their normal gene sisters from becoming deactivated.

If this all seems complicated, that's good, because living things *are* complicated. In computer programming a "kludge" is an ad hoc and perhaps clever but inelegant alteration to a program to accomplish some added purpose, or maybe to fix a bug. A program with many kludges can be complex and difficult for an outsider to decipher. But kludges are how evolution operates. For example, our ancestors needed a tail and we still have the gene for making one; rather than neatly excising it when the need for the tail disappeared, natural selection just turned the gene off.

Although the general ideas of science can often be described succinctly, there is an awesome complexity to biological systems that doesn't come through in such accounts. One might describe the hippocampus as a tiny structure deep in the brain that plays an important role in emotion and long-term memory, and as far as it goes that is quite accurate; but the standard textbook on the hippocampus is several inches thick. Another recent work, an academic article reviewing research on the interneurons—just one type of nerve cell in another part of the brain called the hypothalamus—was over a hundred pages long and cited seven hundred intricate experiments. Few of us would have the patience or the ability to digest such publications, but fortunately for the human body of knowledge, there are those among us, shaped by who knows what interplay between their genome and their environment, who consider them compelling reading.

Being human, we often hope for simple links, like an easy correspondence between a single gene and a trait or a disease, and scientists sometimes find them—as in cystic fibrosis and sickle-cell anemia. Deepak's metaphysics is always free to offer easy but vague answers and unsupported statements such as "You can't start from a meaningless cosmos and get to the rich meaning of human life" or "Human life is embedded in the domain beyond space-time," but science must give answers that are *true,* as determined by experiment, and the truth is rarely simple.

The richness of life comes from its complexity. It is a great gift that one can live and love and function as a being, the cooperative effort of thousands of trillions of cells, intricately and elaborately organized. And yet amidst all life's complexity one can still find unity. I said above that it is only 0.1 percent of our genes that differentiate one human from another. The gene difference between a person and a chimpanzee is only about fifteen times that—we share 98.5 percent of our genes with those primate cousins. And we share over 90 percent with mice, and 60 percent with the lowly fruit fly. There seems to be integrity to life on Earth, resulting from its common basis, the molecule of DNA.

We are all here—from the grape to the fruit fly to the human—

carrying our DNA forward. Every creature on Earth is a unique expression of it. But unique as each one is, all organisms share the same evolutionary mandate: to promulgate their own special version of that extraordinary molecule that—in 1869, in the guise of a being called Friedrich Miescher—made the discovery of its own existence.

DEEPAK

From a spiritual perspective, my role isn't to argue against Leonard's fine account of how genes have evolved into the rich complexity that they display today. In all the great questions that face us, science is our best means of describing physical events. But spiritually speaking, genes exist to do more than provide a recipe book for life. Let's see what that "more" is, which contains many surprises.

I attach great importance to the small number of human genes, but it takes a bit of discussion to show why. As the Human Genome Project was nearing completion in 2003, bets were informally placed. Would it turn out that we possess 80,000 genes, or 120,000? It was assumed, as the most advanced species on the planet, that our complexity required far more genes than any other species. What a shock, then, when the number came in at between 20,000 and 25,000, about the same number as a chicken or a lowly worm like the nematode. Corn had *more* genes, which was baffling. We experienced a minor version of the shock that hit the Victorians when Darwin revealed that *Homo sapiens,* like all mammals, was descended from fish.

In both cases the shock proved highly productive. As Leonard has described so well, inheritance is far more flexible than anyone ever supposed fifty or even twenty years ago. At that time we were getting to the point where "my genes made me do it" was turning into a universal explanation: my genes made me overeat, caused my depression, reduced my sex drive, made me suicidal, or made me a believer in God. The code of life was being interpreted like a code of law. Cells are not fixed structures, however; they are fluid, changing, and dynamic. They respond to thoughts and feelings; they adapt to the environment with all the unpredictability of a person. For anyone who values life's rich possibilities, that's very good news.

When schoolchildren are taught about the double helix, the example used over and over is that there is a gene for blue eyes, another for blond hair, and yet another for freckles. This gives the impression that one gene equals one trait, but that is the exception, not the rule. I mentioned before how frustrating it was for geneticists to discover that what should be a simple link to how tall a child will grow has turned out to be a complex, dynamic process involving not just twenty different genes but a host of outside factors from the environment. Alzheimer's or cancer seems to involve even more genes.

As a result of this murkiness, geneticists eager to fulfill the promise of DNA to improve human life are redoubling their efforts. Since that's also a spiritual goal, how can the two join forces? One way is to quickly get past chemical determinism. The public is still being told that there might be a "criminal gene," for example, that explains antisocial behavior. There's speculation that such a gene could even be offered as a defense in court, and it wouldn't be a big step to propose that antisocial genes could be removed through some kind of medical procedure, say, for the good of the criminal and society as a whole. But as genetics is being forced to abandon the simplistic notion of finding a single gene to fit every disorder, there is an opening for spirituality, which stands for free will, consciousness, creativity, and personal transformation—the opposite of chemical determinism. We should celebrate being released from our genetic shackles, while at the same time seeking more insight into how genes relate to consciousness.

DNA is treated by biologists like any other chemical sequence, but its behavior breaks the rules of mere objects. It spontaneously divides itself in half, turning into two identical versions of itself. It encodes life but also death, since there's a gene for cancer that must be triggered for malignancies to develop. Why in the world would evolution retain such a gene when its whole purpose is to sustain life? And at an even more basic level, how do genes make inanimate chemicals like hydrogen, carbon, and oxygen come to life?

Tracing these issues back to the genome is a feature of materialism. Instead of flying in the face of facts, the spiritual perspective calls for expanded facts. Without them, we can't hope to solve, for example, how DNA deals with time. Genes precisely time their actions years or decades in advance. Baby teeth, puberty, menstruation, male-pattern baldness, the onset of menopause—all these appear on a timetable; the same may also apply to cancer, which is largely a disease of old age. How does a chemical keep track of time? I asked a cell biologist that question and he pointed to telomeres, genetic material that caps the ends of genes like a dangling tail. (We previously touched on them in discussing the nature of time.) Telomeres bring a genetic word to a stop, the way that a period brings this sentence to a stop. But telomeres degrade over time, and aging could be based on their growing shorter and shorter, leading to cellular degradation and higher risk of harmful mutations.

But if the telomere really is like a clock, where did it get its sense of timing? Rocks are worn down by wind and rain, but that doesn't make them clocks. Besides, how can telomeres lead to the harmful effects of aging and also to the beneficial effects of losing your baby teeth and passing into puberty? Even more mysteriously, DNA coordinates many different clocks simultaneously, since the timings of the processes I mentioned are very different from one another. Menopause obeys a clock that takes decades to unfold, while the steady production of enzymes in a cell takes a few hundredths of a second, red blood corpuscles follow a life cycle of a few months, and so on.

The reader will see where this is going. Genes don't behave like ordinary things, because they serve consciousness. Timing requires a mind, and leaving mind out of the equation fatally flaws any genetic theory. To a materialist, the thought of mind outside the body is outlandish, but there is simply too much that mindless, random chemical reactions cannot explain. At bottom, a deep spiritual issue is at stake: free will versus determinism. At first, determinism was just physical, but lately it has been invoked to rule human behavior, too; whether you're acting criminally, depressed, or awed before God, the argument is the same: if genes cause X, and you cannot change the genes that you're born with, then X is here to stay.

Everyday experience belies this logic; none of us feels controlled by the nucleus of our cells. Leonard allows for environmental influence on our genes. I would make it a decisive factor. Identical twins offer a good test case. They are born with the same genes, but as life progresses, twins make different choices and go through different experiences. One twin may run away with the circus while the other joins a convent. One may become an alcoholic while the other becomes a vegan. By age seventy, the expression of their genes will be completely different from the perfect match they displayed at birth. In other words, the chromosomes haven't altered, but the genes that got triggered, along with the products they produce in the tissues, have widely diverged. The escape route from chemical determinism was always there, waiting to be used.

Genes have no effect until they are switched on; they remain mute, as it were. When they do speak, a lifetime of experiences shapes the words expressed, even though the starting point is the same alphabet. Genes don't tell our story; they give us the letters to tell our own story, and that genetic expression can be positive or negative. If twin A habitually lives with low sleep, high stress, a bad diet, and no exercise, such a lifestyle is likely to lead to drastic outcomes compared to those for twin B, who has chosen the opposite lifestyle. Studies in positive lifestyle choices by Dr. Dean Ornish and his research team have shown that more than four hundred genes change their expression in a positive way if someone practices the well-known preventive measures of diet, exercise, stress management, and good sleep.

In a word, the tables have been turned. Where genes used to take responsibility off our shoulders for the things we don't like about ourselves, now they have become the servants of the choices we make. "Soft inheritance" is happening every second, as your cells adapt to the instructions you give them. For decades we've known that depressed people are at higher risk for disease, as are lonely people, the recently widowed, and executives who have been forced out of their jobs. The body can't respond to such traumas without genes being involved, but back when genes were considered fixed, permanent, and unchangeable, no one thought much about the connection between the environment and DNA. ("Environment" in this case is a broad term to cover any outside influence on a cell.) Now doctors routinely warn pregnant mothers that they put their fetuses at risk by smoking and drinking, for instance, since we know that toxic chemicals in the bloodstream degrade the environment of an unborn child.

The next step was to show that toxic behavior can have the same effect. For a long time it was assumed that embryos develop automatically from the blueprint of the DNA inherited from their parents. As long as the fetus received the right nutrients in the womb, the theory went, the blueprint would unfold stage by stage until a baby was born. But as Professor Pathik Wadhwa, a specialist in obstetrics and behavioral science at the University of California, Irvine, puts it, "This view has more or less been completely turned upside down. . . . At each stage of development, the [fetus] uses cues from its environment to decide how best to construct itself within the parameters of its genes."

Suddenly we find that we can add a new chapter to autopoiesis, or self-creation. The unborn embryo is part of a complex feedback loop, assessing the present to create a future for itself. DNA does the same thing. It takes cues from a person's thoughts, moods, diet, and stress levels (to simplify the thousands of chemical signals coming into a cell at any given moment), and based on those cues, it expresses itself. A stressed-out mother passes on higher stress hormones to the fetus. Premature birth is then a risk; so is much else. Professor Wadhwa continues, "The fetus builds itself permanently to deal with this kind of high-stress environment, and once it's born may be at greater risk for a whole bunch of stress-related pathologies."

Where does that leave us? Our knowledge of medicine and biology has been shaken to the core. Genes do not control themselves. They are controlled by the entire mind-body system: in others words, we aren't pawns but masters of our genes, which respond to everything we think and do. The signals from the epigene, the sheath of proteins that surrounds our DNA, are capable of causing thirty thousand different expressions from a single gene. The program of life is dynamic, constantly changing, and under our influence insofar as we make good or bad choices.

More and more, researchers are realizing that genes are more like rheostats than like on-off switches. Areas of "junk DNA" are vitally important, as Leonard touches upon, since they decide which genes to turn on, how much activity a gene expresses, when the activity occurs, and how it relates to thousands of other genes. But as we now know, these genes don't control themselves. No one can tell the final story of the gene until it includes the way in which we metabolize experience. The epigene shows us that even invisible things like stress turn into bodily processes; whatever you feel, every cell in your body also feels. None of this comes as a surprise to those of us who work in the realm of spirituality. The very basis of the spiritual worldview is that *everything* is entangled and interconnected; one process diversifies into thousands of specific processes without losing its wholeness.

I find myself deeply moved when I reread some lines from the great Bengali poet, Rabindranath Tagore, as he addresses his creator. "Time is endless in your hands, my Lord. There is none to count your minutes. Days and nights pass. You know how to wait. Your centuries follow each other perfecting a wild flower." I don't read these words theistically, based on the existence of the God of any particular faith. What moves me is the patience and intricate workings of cosmic intelligence, which moves through us in order to create us, as life unfolds from within itself.

11

Did Darwin Go Wrong?

DEEPAK

Spirituality owes heartfelt thanks to Charles Darwin, although he would be surprised to hear it. When people wake up all around the world with similar aspirations—"I want to better myself, I want to grow, I want to fulfill my potential"—they are taking personal advantage of Darwin's great discovery, evolution. Darwin didn't intend for people to think of their personal evolution, much less their spiritual evolution. A disillusioned theology student who harbored a bitter distrust of the Victorian God—benign, merciful, and a loving father to humankind—Darwin struck the decisive blow against that God. The theory of evolution liberated science from religion, toppled the myth of perfection in Nature, and supplied an airtight mechanism for how each species came into being.

Yet great ideas spread far beyond their discoverer's control. Darwin's blow against perfection was also a blow against sin, the "human stain" that could be atoned for but would always return. Evolution opened a way to escape the trap of sin by offering hope for progress in all aspects of life, although it took a long time for such a humane implication to strike home. At first people seized on another aspect of Darwin's theory: the violent struggle for survival that left only the fittest standing. Alpha-male industrialists could abuse their workers on the grounds that Nature intended the strong to rule over the weak, and tyrants could justify themselves the same way. But today it is in the interest of spirituality to promote evolution over materialism. Where Darwin went wrong was to see evolution as a mindless mechanism.

Spirituality can restore it as a mindful way to make life better, through higher consciousness. *Wake up and evolve.*

It's fascinating to follow the bright-eyed young naturalist as he set sail for South America in 1826 on the HMS *Beagle,* a voyage that would last five years. He dug up the fossilized skulls of giant extinct mammals and wondered how they related to present-day mammals. He pondered why the sea iguanas of the Galápagos Islands, among all iguanas in the world, took to the ocean for their food. He rode on the backs of six-hundred-pound Galápagos tortoises (it's easy to get them going with the tap of a stick, but much harder to keep from slipping off), and he speculated why the shells of tortoises from each island were subtly distinct from those on neighboring islands, having perhaps a larger flare at the bottom, a slightly different color, or a helmet-like extension at the front that covered the beast's head.

These oddments and observations gathered in Darwin's head, and once he got back home to England, his thinking went into a feverish state. After bouts of writing when ideas poured out of him as if on their own, he eventually hit upon the idea that has been called the most brilliant ever conceived: the tree of evolution that links all living things. In Darwin's system, adaptation is the driving force behind evolution. Gazelles have adapted to outrun lions. Clownfish have adapted to hide safely amid the poisonous tentacles of sea anemones. Humans have adapted to use opposable thumbs so that we can make better tools (and weapons). Species change. A single evolutionary tree grows thousands upon thousands of branches, and some die off while others thrive and flower.

Evolution is invoked by eager atheists to grind into dust everything that the word "God" implies. But atheists are fighting yesterday's battles. Today evolution is bringing people closer to God. Darwin devised a perfect physical mechanism only for the life-forms that preceded us. As long as mountain gorillas struggle for food and breeding rights, some will be more successful than others. Dominant males can pass on their genes while submissive males sit by sullenly and envy them. Taller trees will stretch for sunlight while shorter ones wither in their shade. But *Homo sapiens* has evolved beyond mere survival of the fittest. We raise food for each other. We nurse our weak, giving their genes as great a chance of being passed on as the genes of the strong. Darwin's universal mechanism stopped applying to us the moment our species learned to shelter our genes, even recessive ones, from the forces of Nature. Gazelles and clownfish don't tape reminders on the bathroom mirror saying, "Note to self: Remember to evolve today." For them, evolution is automatic. That's no longer true for us.

Spirituality can be seen as a higher form of evolution, best described as "metabiological"—beyond biology. We have been on this track for at least 200,000 years. Our ancestors, such as Neanderthal man and *Homo erectus,* were preparing the way as far back as 1.8 million years ago. When they chipped stone axes out of flint, our ancestors thought out what they were doing. Once you wake up with a desire to do something besides eat and breed, you start choosing X over Y. Conscious decision making turns the future into a series of choices. Neanderthals were advanced enough to place their dead in cave tombs, and some evidence suggests that the deceased were decorated with ornaments. Beauty, it seems, had become a choice, too, along with reverence, and perhaps even a sense of the sacred.

Yet modern Darwinists act as if humans are still in the primal state of Nature. Not that primal was ever simple. Survival is complex—an intricate tapestry—even among lower creatures. Penguins have been swimmers instead of fliers for over 36 million years. Diving for fish was a spectacular success in evolutionary terms, although the original penguin was feathered in brown or gray (this was discovered by examining fossilized pigment cells). Why did that change to the black-and-white penguin suit we smile at today? Darwinism has only one answer: competitive advantage. The original penguin was five feet tall and weighed twice as much as the present-day emperor penguin. Why did penguins grow smaller? That, too, must have contributed to survival. Darwinism is forced to explain any change the same way, because it cannot get past its one-eyed focus on the struggle for food and breeding.

But species don't just compete for survival; they also cooperate in a relationship known as mutualism. Bizarre tubeworms that live near hydrothermal vents at the bottom of the sea survive with no guts, thanks to bacteria that provide a digestive function in return for the hydrogen sulfide or methane that the tubeworm provides for them. The clownfish I mentioned earlier has developed an adaptive mucus to protect it from the poison in a sea anemone's tentacles. Using those tentacles as a safe haven from predators, the territorial clownfish returns the favor by protecting the sea anemone from sea-anemone-eating fish. To say that competition alone drives evolution is clearly wrong on the face of it.

The same goes for the so-called selfish gene. Genetic theory had to come up with an answer for why evolution sometimes favors death over life. Survival is not always a creature's sole drive. Honeybees are equipped with a stinger to protect the hive, but when they use the stinger, it pulls out, fatally injuring the bee. You cannot explain this kind of self-sacrifice as contributing to survival; the bee is dead. So evolutionists had to back up a step. It's the honeybee gene that is fighting for survival, not the individual insect. Specifically, the genes in the queen bee must survive, which means that lower-ranking bees can sacrifice their own lives so long as the hive as a whole benefits. The same argument applies to female spiders that bite off the heads of the male during mating, or by extension to the millions of fish eggs that drift through the sea providing food for other fish without ever having a chance to hatch. If a hundred hatchlings survive while a million perish, the gene pool continues.

As a credible explanation, the selfish gene borders on the absurd. It doesn't get us to the real locus of evolutionary change, the intelligent cell. DNA cannot control how a gene responds to the environment, for example, because DNA is deaf, dumb, and blind. It sits passively inside the nucleus of a cell; it replicates as RNA to produce the enzymes and proteins for cell growth. Nowhere in this chain of chemical events is there a way for the gene to look out upon the world and decide to be selfish or unselfish. The only valid way to explain self-sacrifice is by inserting the one element that materialists abhor: consciousness.

A honeybee can serve the hive when there is an overriding purpose—to keep the whole alive despite the death of some parts. The human body clearly preserves the whole over the parts. White blood cells, for example, die after they consume invading bacteria. Every cell in the body has a programmed life span, from a few weeks in the case of skin and stomach cells, to the lifetime of the body itself in the case of some brain cells. The mindful principle that the whole is more important than its parts extends to our entire planet. The purpose of ecology is to maintain itself, not any one plant or animal. Yet within this scheme hundreds of thousands of species can thrive at the same time, even those that are mortal enemies.

A mindless mechanism will always be insufficient to explain how life evolves and thrives. There are too many opposites, like competition and cooperation, selfishness and altruism, that coexist. Conscious choices are being made throughout Nature. It's not just the critics of Darwinism who found flaws in the theory. Today as many as eleven reinterpretations and revisions are competing for primacy among evolutionists themselves (in classic Darwinian fashion). Each revision tries to fill in a gap or correct a mistake. Progressive Darwinists, for example, try to explain how infinite variety develops from limited genetic material. Human beings have only twenty-three thousand genes, of which 65 percent are so basic that we share them with a banana. These progressive Darwinists look more closely at the developmental stages of growth—hence their nickname of "evo devos"—and they have discovered that stretches of seemingly random sequences in our DNA are helping to turn genes off and on, acting as "molecular fingers" controlling a bank of switches so that embryos in the womb can develop along entirely unique lines.

Another camp, the collectivists, recognized that evolution required cooperation as much as competition. They focused on how the enormous leap from one-celled organisms to eukaryotes, or multicelled organisms, was the result of a cooperative venture with plants, who had developed photosynthesis. Strict Darwinists had reason to resist, because cooperation defies the notion of the selfish gene, and only after a twenty-year struggle did cooperation become accepted as the basis of life.

Other camps snip off other pieces of the puzzle to solve. The complexity theorists study how a system can become so intricate that it spontaneously gives rise to ever greater complexity. Without that ability, a single fertilized ovum couldn't develop into fifty trillion cells—our best estimate of the cell count in an average adult human. The so-called directionalists tackle the way that complexity and cooperation never stop—two kinds of one-celled organisms cooperating two billion years ago has snowballed into a planet where every living creature affects every other. Seven other specialized camps are busy injecting bioengineering, design, God, and metaphysics into the scheme to see if any fit. All the parts of this patchwork are aimed at pinpointing exactly how the mechanism of evolution works.

What if you look at the whole picture at once? Because billions of living parts are involved, the whole is nearly impossible to glimpse, but one can see that all of life is evolving, here and now. It's time to adopt a holistic approach to evolution, and no better case exists than our own species. Early hominids like Lucy, roaming the African grasslands 4 million years ago, evolved into humans like *Homo erectus* about 1.8 million years ago. *Homo erectus* looked incredibly like us. It was well over five feet tall (whereas Lucy was under four feet). It had lost the fangs of primates such as chimpanzees; its hips had widened; it walked upright all the time instead of sometimes crawling or climbing in trees; it had lost almost all its body hair; and sweat glands had replaced panting through the tongue as a way to cool down. (A body that can cool down is able to run long distances after prey, which early man had to do since he couldn't outfight large animals; present-day bushmen of the Kalahari Desert continue to chase antelope for hours at a time until the animal drops from fatigue and is easily dispatched.) Larger brains developed outside the womb, after a child was born (this was necessary because a fully formed human brain cannot pass through the birth canal). It is hard to believe that each of these adaptations dripped into the hominid gene pool as a random event. The arrival of *Homo erectus* looks purposeful and holistic.

But where does purpose originate from? Intelligence seems to guide structure. Some anthropologists speculate that *Homo erectus* took a great leap in more than physical traits. As a primitive toolmaker he learned to judge which flints made good blades and which didn't. That implies the capacity of reason. To ward off predators at night *Homo erectus* might have tamed fire twice as early as the 750,000 years ago that is currently accepted. Studies of brain shape indicate that the first humans may have had much the same language centers that we do: so did they speak? As one speculation bounces off another, it seems likely that multiple traits appeared at nearly the same time, rather than single traits at random. Each change provided a catalyst for others. Standing upright allowed for long-distance running, which allowed for more food, which allowed for a larger brain (the most calorie-hungry organ of the body), which allowed for the higher reasoning necessary to discover fire and take care of helpless babies while their brains matured.

Beyond Darwinism lies a better way to view life on our planet: intelligent feedback loops. Life creates a new trait, gets good at it, and watches itself as it gets good. Such a feedback loop isn't mindless; it has purpose, desire, and intention. For example, every person has a sense of balance. It is innate, a given we don't need to think about. You can improve upon it, as people do when they learn to ski, skateboard, or walk a tightrope. When you look carefully at what is going on when a beginner is learning to ski, to all appearances there's a lot of falling down and flailing. But this chaotic behavior isn't what it seems. Every mistake is contributing to a feedback loop inside the brain that is learning, step by step, to master a new skill. Behavior that appears to be random actually is serving a purpose, even though you cannot observe the purpose by just watching the random events.

If you keep trying to ski, you will train your sense of balance even further. In a word, you are causing it to evolve. The whole body joins in the enterprise. Your long muscles adjust as you lean one way or another. Your ankles adapt to the stiff ski boot; your breathing changes as you focus intently. Your eyes feed information to your brain about how the slope looks as it races by. None of this activity is isolated; it is all funneled by your single-minded intention. And although skiing is a new development, you have possessed the potential to learn it from birth.

What applies to your use of intelligent feedback on a ski slope can be extended everywhere in Nature. Darwinism is stuck if it insists that every trait came about as a result of getting better at finding food and a mate. Creatures gain an identity, which discovers itself through intelligent feedback loops. Horses learn to be better horses, snakes to be better snakes. Each is a special, unique set of qualities that mesh beautifully. The mistake we make is to humanize such intelligence. Evolution doesn't need to employ a complex brain. Feedback loops are universal. One-celled animals use them, too, since even the most primitive creature orchestrates eating, breathing, cell division, and motion.

Spirituality restores purpose and direction to their rightful places at the heart of evolution. As humans we know where we want to go (at least we hope we do), and our intentions have led to a world where atom bombs coexist with peace conferences, cars with pedestrians, lumber companies with conservationists. We have entangled ourselves in a web of desires, some tending to make life better and others tending toward self-destruction. If we want to evolve beyond our worst impulses, the only way is through a higher purpose that benefits everyone. Religion tried to supply that higher purpose through God, but as we see in holy wars, sectarian violence, and terrorism, God can serve destruction, too. This is why spirituality, the taproot of religion, is our last best hope. It holds out the possibility for the evolution of consciousness.

Darwinism (as opposed to Darwin himself) stands as a huge obstacle for saving us, which is deeply ironic but undeniable. Evolutionary theory is used to support the following false ideas:

Life is completely physical.
Evolution proceeds by accidental mutations.
Mind and higher purpose are illusions.
Survival is the ultimate goal of all living things.
Competition is the driving force in Nature.

Darwin himself isn't to blame for these notions; his goal was purely to show how one species gives rise to another. He didn't invent the phrase "survival of the fittest," much less the gloomy Victorian view of "Nature red in tooth and claw." But seeds were planted through Darwin's aversion to God and his focus on mechanism. His followers and descendants grew those seeds into a theory where randomness and mindlessness prevail. As long as this network of ideas colors your worldview, there is no reason to believe that consciousness can evolve. Remove those false assumptions, however, and it becomes clear that consciousness has been evolving since the very beginning, and it will never stop.

LEONARD

Deepak pleads passionately about the need for humanity to evolve beyond its worst impulses, and argues that it can do so through a higher purpose that benefits everyone. He is right that religion has often failed to provide that, providing instead motivation for conflict and destruction. And I believe he is also right in saying that we can rise above the basest of the survival-of-the-fittest mechanisms because the social and altruistic behaviors that distinguish us from other animals are also a product of evolution, thus part of our very nature, as I will discuss below. It is those behaviors that can enable us to find salvation from the many dangers we now face. Deepak's spiritual approach can serve that end, too, especially if it encourages us to express our innate altruism, or nurtures culturally based altruism. But we must be careful not to allow ideas about what we must do to improve human life to influence what we believe *is* human life.

Deepak tells us that spirituality owes a deep thanks to Charles Darwin, but the picture he paints of Darwin's ideas today is a portrait of a theory racked by confusion and chaos. "As a credible explanation, the selfish gene borders on the absurd," he writes, and "it's not just the critics of Darwinism who found flaws in the theory. Today as many as eleven reinterpretations and revisions are competing for primacy among evolutionists themselves (in classic Darwinian fashion). Each revision tries to fill in a gap or correct a mistake."

Did Darwin go wrong? Are scientists really climbing over one another to plug holes in the bow of a sinking ship, or to claw their way onto a lifeboat?

The answer is absolutely not. With the exception of a handful of creationists motivated by their religious beliefs, no scientist doubts the basic idea of Darwinian evolution, or that natural selection is the mechanism behind it. That's why working scientists don't call themselves "evolutionists" or "Darwinists." These terms are common among creationists (from whom Deepak rightly wishes to distance himself), because to use them gives the mistaken impression that among biologists there are some who believe in evolution and others who don't. Calling a biologist an "evolutionist" or "Darwinist" is like calling a physicist a "round Earther" or a "Columbus-ist." The original idea of a "round Earth," dating back to the ancient Greeks, held that the Earth is perfectly spherical. The flat-Earth theory resurfaced from time to time until Columbus made his famous voyage, which provided dramatic evidence for the round-Earth theory. Still, over the years there were those who made "revisions"—people like Isaac Newton—who realized that the Earth is not really a sphere. They "reinterpreted" the round-Earth theory, predicting and measuring the Earth's slightly squashed shape, and studying its details, causes, and implications. Does the need for revision and reinterpretation mean we should revert to the flat-Earth theory? Of course not. But filling in the "mistake" or "gap" in the theory did not discredit the idea that the Earth is round, and physicists today would roll their eyes at anyone worried about falling off. Similarly, there are debates about the relative contributions to natural selection made by genes, individuals, or groups of individuals, and it's true that understanding the detailed patterns of evolution in different species is complicated, but the basic idea of natural selection, and the fundamental role of randomness in the process, is not in question.

What should we make, then, of all the biologists studying different aspects of evolution? Deepak calls them camps, and remarks that scientific ideas about evolution are themselves competing in "classic Darwinian fashion." The comment sounds damning, as though there is a war going on, which may ultimately remove Darwin from his place of honor in the scientific pantheon. But this is just the normal scientific discussion that surrounds every theory. In fact, it illuminates an important difference between science and metaphysics. In metaphysics one has the luxury of embracing any attractive idea. In science new ideas can be incorporated into theories—as happened in the round-Earth example—but the only new ideas that survive are those that experimental evidence shows to be valid. It is one thing to say that "the selfish gene doesn't hold up as a credible explanation," but it is quite another to prove it.

What does it take to "prove" something in science? One of course wants to test a theory's obvious predictions, and to collect evidence that it explains what it claims to explain. But that is only the beginning. In fact, more important than gathering evidence that a theory is right—and more exciting to a scientist—is trying to find situations in which a theory's predictions might be wrong. Scientists are like devil's advocates—or your annoying little brother; they question everything, eager to concoct an exceptional situation that proves you are misguided. That's not a flaw in the fundamental character of science; to the contrary, it is how science makes progress. So when scientists say they've found evidence in support of a theory, they often mean they have been looking for a new way to challenge the theory, and the theory passed the challenge. This happens even in well-established theories like evolution, but it should not be interpreted as a sign that the theory is in trouble.

Take Newton's law of gravity, for example, which accurately describes, under the conditions of everyday life, the force of gravitational attraction between objects. Experimental physicists are still testing that law, though in the three hundred–plus years since Newton proposed it, no one has ever found a deviation, except in extraordinary circumstances, such as those described in chapter 2. So why are scientists still looking to poke holes? Because in the centuries since Newton, scientists have been able to verify only that Newton's law of gravity correctly describes the attraction of objects at distances ranging from a few thousandths of a centimeter to a light-year, but new experimental methods now allow scientists to test it at even shorter distances, and it would be a discovery with exciting implications if the law were found not to hold at all distances. That is valid science, but it is not an indication that physicists are abandoning the theory.

What if a theory does fail an experimental test? That means the theory must be altered, but it doesn't necessarily mean its basic principles are wrong. The round-Earth theory is a simple example—the Earth is not perfectly "round," but though the details of the theory changed as we learned more about the Earth's shape, the main idea that the Earth is not flat survived. Genetics, as we have seen, has also evolved from the simple early models that arose when the structure of DNA was first revealed to the very complex reality scientists have uncovered in the decades since. Though a theory can often be succinctly summarized, the headline telegraphing its meaning usually belies considerable complexity, both in the concept and in its application to situations in the real world. Much of the work of scientists concerns understanding the details of that complexity, and adjusting or elaborating on the theory as we keep learning more, as was done in the theories I've just mentioned.

In criticizing Darwin, Deepak focused on a facet of the theory of evolution that is relevant to his humanitarian goal, and something he believes Darwin's theory cannot explain: cooperation among individuals, which seems to contradict the idea of selection through competition. I agree that this is an important challenge for evolution, one of those critical blanks that must be filled in. Darwin himself wrote that it is "by far the most serious special difficulty, which my theory has encountered." Darwin believed the answer was that the community benefits, that natural selection in this case is operating on the level of the group, rather than the individual. As we'll see, there is a lot more to it than that, but there *is* an answer, and the work of filling in that blank was nothing above and beyond the analogous filling in that occurs in all theories, from the round-Earth theory to the theories of electromagnetism and the quantum that are responsible for most of modern technology.

Deepak wrote that Darwinian evolution must be wrong because, if it were correct, "competition and cooperation, selfishness and altruism" cannot coexist. It is true that the headlines of evolution theory—natural selection through competition, and survival of the fittest—seem to disallow cooperation; but as often happens in science, it turns out that if you read the whole story, you get a far more nuanced picture, and in this case a surprisingly wonderful one—of the kind that even Deepak would welcome.

Einstein is said to have remarked that everything should be made as simple as possible, but not simpler, and in addressing this issue, I will try to keep my toes on that fine line. Can competition and cooperation, selfishness and altruism, coexist? Richard Dawkins, who thirty-five years ago coined the term "selfish gene" in his book of that name, now says he has second thoughts about the term because it can be misleading. There is indeed a problem in that the book's title is now widely quoted even though most people have not, in Dawkins's words, read the rather "large footnote of the book itself." A good alternative title, he now offers, would have been "The Cooperative Gene." This seems odd, that a gene can be described as both cooperative and selfish. Let's see why he says that.

Consider Deepak's example of kamikaze bees. They belong to an order of insects called Hymenoptera, which also includes ants and wasps, the social organisms I described earlier. Such insects are famous for their apparent altruism and cooperative behavior. In these insects, the society as a whole is like an organism. The majority of individuals are sterile workers. Some ants tend to the nest, others to battle, others to food. Among bees, intruders are recognized and attacked, with individuals playing the role of the cells in our immune system; and together the metabolism of individual bees regulates the temperature in the hive nearly as well as the individual human body regulates its own temperature—even though bees are not "warm-blooded." In each hymenopteran colony there are also a minority of individuals (typically one of each sex) that reproduce—the female queens and male drones—and it is through those insects that the gene line flows. In the advanced societies, the queens and drones do nothing but reproduce, while all food, defense, and nanny tasks are taken care of by the workers. Each female hymenopteran has the genes to become any type of worker, even the queen. But as we saw in my last chapter, the type of genes that are turned on can depend on the environment, and in this case the environment—especially the food provided—determines whether a female develops into a particular type of worker, or a queen.

Given this social structure, the kamikaze behavior of worker bees who die after stinging makes perfect evolutionary sense, because it does not diminish the survival of their genes—worker bees never bear offspring—while it does enhance survival of the hive, and hence of the bees that do reproduce. As Dawkins writes, "The death of a single sterile worker bee is no more serious to its genes than is the shedding of a leaf in autumn to the genes of a tree."

But there is still an important question: why did the workers' reproductive ability wither, like an unused appendage? Can it really be that it is somehow more efficient for worker bees to pass their genes along by aiding the reproduction of the queen—their mother—than by having offspring of their own? The answer is astonishing. In most animals (except in the case of identical twins), a female is more closely related—that is, genetically more similar—to its offspring than to its sisters. But when scientists examined the reproductive process of hymenopterans, they found something quite odd. As a result of particular quirks in the bees' reproduction, a female is genetically closer to its full sisters than to its children of either sex. A gene encouraging sacrifice for the good of the hive, which would aid in the creation of sister bees, is therefore favored by evolution over a gene for making offspring directly, and so the fertility of the worker bees became genetically irrelevant and disappeared. Kamikaze bees look altruistic, but their behavior is in their genes' best interest!

There are many other details to the story, as usual. For one, though female hymenopterans are closely related to their sisters, they are not as closely related to the males, and so if the system I described works, one should expect there to be many more female offspring than males. It is even possible to predict the optimal sex ratio, and this turns out to be very close to what is observed. Another detail is that there are some species of social insects in which a queen mates with multiple males, resulting in sisters that are not full sisters—that is, not as closely related—yet these societies exhibit the same altruistic behavior. That mystery was finally explained by a striking study in 2008 in which sophisticated DNA analysis showed that when, millions of years ago, the current social structure of social insects evolved, the queens in all lineages were monogamous, and sister bees *were* all very closely related. Cooperation among social insects, a challenge evolution had to answer, has turned out not to represent evidence of a flaw in the theory, but rather to provide convincing support that it is correct.

Associations of mutual benefit also occur in animals other than social insects. But there are limits to altruism. Consider the case of an animal that would give away food if it had plenty, and another animal that was near starvation. The chances of the altruistic animal avoiding starvation would diminish just a little, while the chances of the other surviving might increase a great deal. But unless the organism on the receiving end shared the donor's genes, the donor would slightly decrease the odds of its living to pass its genes on to the next generation, while its genes would receive no survival benefit in return. Such an animal would have chronically a little less to eat than a selfish cohort that took but never gave. As a result, according to natural selection, animals with genes bestowing this kind of blanket altruism should be expected to die off—but if an altruist is choosy regarding the animals it shares with, things change, and we see that kind of altruism in many species.

One way to be choosy is to have the sophisticated ability to recognize and remember who returns the favor, and to stop sharing with individuals who don't. Animals of this sort help others in times of need, but in exchange receive help when they are in need. That is called reciprocal altruism. We all have some tendency to practice it, and behavioral economists have studied that tendency in great detail, setting up games in which volunteers cooperate and compete for monetary awards.

A more selfless style of biological choosiness is to share only with relatives—a type of altruism called kin selection. When an organism shares with relatives, especially close relatives, there is a good probability that the recipients of the kindness share its genes. As a result, though an organism might reduce its own chances of survival slightly by sharing, when it boosts the chances of the relative, it increases the odds of its own genes' survival. The net result of such acts can be that the altruism gene is likely to be passed along, so this kind of altruism tends to survive. Kin selection has testable consequences. For example, it predicts that altruism in the animal world is more likely toward relatives than toward unrelated animals, and that the closer the relationship, the higher the degree of altruism, predictions that have been confirmed in empirical work on species ranging from birds to Japanese macaque monkeys.

Darwin wasn't wrong, but as Deepak says, Darwin takes us only so far. Most people, when stepping into a street today—even a deserted one—will look both ways, often without even thinking about it. We have genes for abilities to detect danger, but there is nothing in our genes that makes us look before crossing the street. We needn't develop a genetic mechanism for that, because each generation can easily solve that kind of problem anew, and the knowledge can be passed down through culture.

The evolution of culture is perhaps more important than genetic evolution to humanity today. Humans have lived in countless civilizations, but the few hundred generations since the ancient Greeks have not been enough for natural genetic evolution to have had much of an impact on us. It's not that we haven't changed—we have; but what most distinguishes us from the civilizations of the past few thousand years is not the effect of shifting genes, but the effect of shifting culture. Stephen Jay Gould noted that in other mammalian species, the "murder" rate is far higher than in human cities. In this and other ways our cultures can allow us to rise above our genetic makeup. That's a key to our survival because, due to rapid technological progress, the environment in which we function has changed drastically over the past centuries. Today's technology brings us great good, but today both groups and individuals have the power to do great harm, either through bad intention (terrorism) or simply through inattention to technology's harmful effects (pollution and global warming). Our best hope for a better future, then, is through the development of values that encourage caring for one another, cherishing knowledge and learning, preserving natural resources, and minimizing harm to our environment. It is only this kind of evolution, which is cultural rather than biological in nature, that can save us.

PART FOUR

MIND AND BRAIN

12

What Is the Connection Between Mind and Brain?

LEONARD

When it comes to sensations, emotions, and the ultimate question of consciousness, science still can't explain the connection between neural patterns and the mind. We can characterize many emotions according to the physiological reactions that accompany them—a blush or change in the electrical conductance of your skin, for example; and we've also made progress in understanding what is going on in your brain, both anatomically and chemically, as you experience those emotions. So we understand a lot about how the brain functions. What we understand very little about is the subjective experience of those emotions, the "felt quality" of experience, as philosopher David Chalmers calls it.

What does it mean to "feel bad," or to experience a burn, or the color blue, or sexual desire?

In 1915 a scientist named Alfred Sturtevant carefully observed what we think of as stereotypical barroom behavior—a couple of males fighting over a female, charging each other, ending up in a chaotic tussle. What made his study noteworthy was that the vertices of this love triangle were fruit flies. Even simpler creatures like nematodes, many species of which are microscopic, also exhibit special behaviors related to mating. Nematodes procreate like crazy—grab a handful of soil humus, and the chances are you'll have thousands of these prolific roundworms within your grasp. So forget trying to grasp the complexities of the human mind—what does sex "feel like" to a creature of the phylum Nematoda? It might seem silly to ask about feelings in a species so simple it can survive being frozen in liquid nitrogen. But for one nematode species, *C. elegans,* we have the complete blueprint of its construction—a map of all of its 959 cells, including the wiring of its 302-node neural network (you can find it online)—and there was the hope that the blueprint would help us to understand how sensations arise from its networks of neurons. Alas, even in a creature this simple, it did not.

What is the nature of inner experience, and how can it be the result of neural processes? How do neural processes create the mind? Chalmers termed that "the hard problem." It's so hard that philosophers and poets, theologians, scientists, and physicists have been wrestling with the question of the connection between the material and the immaterial worlds for millennia.

Plato, for example, viewed people as having an immortal soul inside a mortal body. Christianity embraced that idea, as did many other faiths, and some early scientists also embraced it. The great seventeenth-century physicist, mathematician, and philosopher René Descartes, like many before him, differentiated between physical substance and mental substance. In his view, the brain was a physical structure, a machine, but the mind—our thoughts and consciousness—was something altogether different, which did not operate according to the laws of physics. Today we call that idea "mind-body dualism."

For Descartes, as for Deepak, it was philosophical considerations that seemed to drive him. In part Descartes was trying to refute "irreligious" people who put their faith solely in mathematics, and would not accept the immortality of the soul unless it could be mathematically and scientifically demonstrated. But Descartes was also grappling with the problem of how to account for physical phenomena in a way that was consistent with his underlying worldview. In this he differed from the Aristotelian tradition, which was the reigning philosophical belief at the time. The Aristotelian worldview holds, as Deepak does, that there is purpose in the universe. In Aristotle's version of purpose, all objects in nature, both animate and inanimate, behave as they do for the sake of some end or goal, sometimes called a "final cause." For example, a stone tossed into the air would be said to fall back to the Earth because it is striving to reach the Earth's center. Unlike most scientists and scholars of his era, Descartes opposed this idea, and its apparent implication that stones can have knowledge of a goal, and of how to attain it. Instead, Descartes took a mechanistic approach, maintaining that nonhuman objects follow physical laws. His theory of mind-body dualism was in part an attempt to dissuade people from assigning mental properties to inanimate objects and nonhuman animals, and thereby to distinguish the human world, which he did see as being guided, ultimately, by mind and purpose, from the inanimate and nonhuman.

Descartes was aware of certain difficulties that plague mind-body dualism from the scientific viewpoint. For example, through what physical mechanism does the mind control the brain? An accomplished anatomist, Descartes eventually came to the conclusion that the interface between mind and brain was a physical structure called the pineal gland, tucked deep between the two hemispheres of the brain. As it is one of the only structures of the brain that does not exist in two mirror-image parts, one in the left hemisphere and one in the right, Descartes thought it was where mind and brain communicated, and he called it "the principal seat of the soul."

Descartes's anatomically grounded theory is not accepted today, even by those who believe in mind-body dualism. The "hard problem"—the question of where inner experience comes from—remains unsolved. But scientists feel no shame in not yet having arrived at the answers. They may come in the next century, or in the next millennium. Or if they are too complex for human understanding, they may never come. In any case, even on the basis of our limited knowledge today, it is difficult to maintain the distinction between an immaterial mind and a material brain. For one, if a realm that obeys physical laws were to interact with a realm that doesn't, wouldn't the interaction cause noticeable exceptions to the laws of nature in the physical realm? Today we can routinely measure physical phenomena, including those inside living human brains, to enormous degrees of accuracy, but we have seen no evidence of such exceptions. If they do exist, why don't we see them? On the other hand, evidence that thoughts and even subjective feelings *are* manifestations of the physical state of connected neurons abounds.

For example, in the course of treating epilepsy patients, neurosurgeons sometimes implant tiny electrodes in their brains and stimulate the tissue with brief pulses of electrical current. What they observe goes far beyond the mechanical responses high school biology students used to observe when they applied electricity to make a frog's leg twitch. Depending on where they place the electrode, the surgeons can cause patients to hear identifiable sounds, like a doorbell or the chirping of birds (when there are no such sounds in the vicinity); to suddenly recall an event from childhood; or to feel urges, such as the desire to move an arm or a leg. These feelings and experiences, which I think we would all agree occur in the "mind," can be traced directly to the physical stimulation of the brain, persuasive evidence that the brain controls the experiences of mind, and not vice versa.

Even more dramatic evidence comes from patients with epilepsy so severe that to bring relief surgeons sever a nerve bundle called the corpus callosum. Such patients are called "split-brain" patients because severing the corpus callosum divides the brain into its two nearly mirror-image hemispheres, with nothing to connect them. Without the corpus callosum bridge between them, the left and right hemispheres can for the most part no longer communicate, coordinate, or integrate information. What does dividing the *brain* in two do to a patient's *mind*? If the mind exists in an immaterial realm, the surgery should not affect it. But if the mind arises solely from the physical brain, splitting the brain should also split the mind.

Neuroscientist Christof Koch wrote about one such case, a split-brain patient who was asked how many seizures she had recently experienced. Her right hand went up, showing two fingers. Then her left hand, controlled by her brain's opposite hemisphere, reached over and forced the fingers on her right hand down. After a pause, her right hand went back up and indicated three, but her left hand went up and indicated only one. The patient seemed to be of two minds, and they were having a spat. Eventually the patient complained verbally that her maverick left hand often "did things on its own." Language, it turns out, is one of the few functions that resides on just one side of the brain, usually the left side, which controls the right hand. But though her right hemisphere could not speak, it could *hear* the remark. Apparently it didn't like what it heard, because at that point a fight broke out between the two hands. If the mind were not reducible to the brain, there is no reason that splitting the brain into two should also split a single conscious mind into, as Koch wrote, "two conscious minds in one skull."

Deepak writes, "It doesn't matter if you track a brain cell back to the atoms that make it up, then farther back to subatomic particles. . . . No one can point to a specific physical process and say, 'Aha, that's where thinking comes from.'" Though it's true that we still have a lot to learn about the connection between our neurons and our thoughts, not knowing "where thinking comes from" does not prove that the source of thought lies in an immaterial realm. Scientists don't deny what seems special about human experience, but they try to avoid explanations of it that are contrary to the evidence. There are currently an estimated fifty thousand scientists worldwide studying the brain, and none of them, nor any of their predecessors, has ever found credible, replicable scientific evidence that people's mental experiences are the result of anything other than physical processes that obey the same laws as every other assemblage of molecules.

That the origin of mind lies in the physical substance of the brain has been repeatedly demonstrated in biology, but it is also demanded by physics. It is of course obvious that if some immaterial entity from another realm knocks a lamp off a table, the laws of physics have been violated. You don't need to study Newtonian mechanics to know that natural law doesn't allow things to jump around without a physical cause. But the immaterial mind, as envisioned by Deepak, doesn't go tossing lamps off tables. Deepak sees it as being a more subtle mover and shaker. And yet one of its chief activities is actually not subtle at all: the immaterial mind, according to Deepak, processes knowledge. In his view, it is this nonphysical mind that is the essence of who we are; it knows what we know, feels what we feel, and makes our judgments and decisions. But according to the laws of physics, the existence of knowledge, thoughts, feelings, or any other kind of information in an immaterial mind—that is, in a realm that has no physical substance—is an impossibility.

The kind of trouble one can run into if one allows for the existence of immaterial information is illustrated by a famous thought experiment conceived by physicist James Clerk Maxwell in 1867. Imagine, as we did in chapter 8, a box of gas with a partition down the middle. This time instead of a hole in the partition, imagine a tiny door in it—a door so small it can be opened and closed without expending an appreciable amount of energy. When the door is shut, the molecules on either side are in a constant state of motion, bouncing off the partition as well as off the walls of the box, but always remaining on the side of the box on which they started. Next, picture a creature, also of insignificant size and mass, standing at the door, observing the molecules and letting them pass one way or the other at his whim. As Maxwell imagined it, this creature has free will and intelligence, but negligible substance. In other words, it resides in an immaterial realm, just as Deepak believes our consciousness does. William Thomson, a contemporary of Maxwell's, nicknamed it "Maxwell's Demon."

Suppose this Demon decides to let only fast-moving, high-energy molecules transit from left to right, and only slow-moving, low-energy molecules move from right to left. Since the temperature of the gas is a measure of the speed of its molecules, over time the gas on the right side of the box will become hot, and the gas on the left side, cool. In chapter 8, I explained why gas molecules in a box will never spontaneously gather on one side, but one can equally well say that they will never sort themselves into hot and cold. If such a scenario were really possible, it would be revolutionary. For example, you could use the temperature differential to drive an engine, which means you'd be able to power a vehicle without consuming any fuel. But that would violate the second law of thermodynamics, which dictates that the entropy—or disorder—of a closed system never decreases. The entropy of the gases in Maxwell's box, however, *is* decreasing, as the Demon arranges them in such a well-ordered fashion.

This violation of the second law, which leaves the physicist wondering where the missing entropy could have gone, occurs because the Demon has been posited as having an immaterial mind. If, on the other hand, the Demon's mind has a material basis, then the "closed system" I described would include not just the box of gas, but also the Demon's mind. Let's look at how that would change the entropy equation. In order for the Demon to do its work, it has to note and remember information about the velocity of the molecules. As that information accumulates in the Demon's mind (or in a notebook, or in a computer's memory if the Demon is a robot), the mind's entropy increases. To understand why, compare an empty room to a room containing furniture. However you arrange the tables, chairs, and other odds and ends, the room will not be as orderly as when there is simply nothing in it. The tables and chairs are like the bits of information cluttering the Demon's mind: as you add information, you increase entropy. The end result: the decrease in the entropy of the gas molecules in the box is offset by the increase in entropy caused by the information buildup in the Demon's physical mind. With that, we understand where the missing entropy went, and we find that the second law has not been violated. (To those clever readers wondering why one can't simply periodically erase the Demon's memory, it turns out that all that does is transfer the entropy elsewhere through the erasing process!)

Physics defines not just knowledge of the kind the Demon possesses as information, but all our ideas, memories, thoughts, and feelings, which means, according to the laws of physics, that they must reside somewhere in the physical universe—whether embodied in the neural patterns in our brains, encoded in a computer circuit, or printed as letters on a page. Even our experiences of beauty, hope, love, and pain arise in a brain that obeys the ordinary laws of physics. Unfortunately, accepting that a mind that harbors information cannot exist in some immaterial realm does not mean that we understand the workings of consciousness. The challenges we face in trying to understand how a neural system that obeys the ordinary laws of physics can give rise to subjective experience make this one of the great scientific projects of our time. Although Deepak would probably call the attempt to locate the mind in the material world a reductionist's pointless dream, many scientists are at work on just that project, complex and impossible as it may seem. And they are making real progress.

Koch wrote that when he started doing research on the question of consciousness in the late 1980s, it was practically considered a sign of cognitive decline—ill-advised as a career path for a young professor, and likely to make graduate students roll their eyes. But he and a few others did work on it, and today those attitudes have changed. There is a whole new science of consciousness. It is legitimate science, and it has helped us understand which structures in the brain produce emotions, sensations, and thoughts, and how they are chemically regulated and electrically connected. We still aren't close to discovering the basis of "mind," or consciousness, as an emergent phenomenon based on interactions among neurons. But every day more evidence emerges to support the idea that mental experiences like beauty, love, hope, and pain are produced by the physical brain. Researchers in Koch's lab, for instance, have developed a way for subjects to activate individual nerve cells deep inside their own brains—concept cells like those I mentioned in chapter 1—enabling them to control the content of an image on an external computer screen by simply thinking about the image they want to see. Experiments like this one, and work that is being done in many other settings around the globe, encourage us to think we are on the right path, though we are far nearer to the beginning of the road than to the end of it.

DEEPAK

Some years ago black colleges in America wanted to raise much-needed funds, and they came up with a brilliant ad campaign. Its slogan was "A mind is a terrible thing to waste." It would be even more terrible to throw the mind away entirely. Leonard does that when he claims that love is understandable as essentially a brain process. This would be a bizarre statement in any case, although it seems somewhat more reasonable when the exact phrasing is pieced back together: "Beauty, hope, love, and pain arise in a brain that obeys the ordinary laws of physics." Love and beauty are core experiences in spiritual life. We need to get to the bottom of where they come from. There is an answer, but to accept it, you have to see the difference between love and the products of a chemistry set.

Leonard calls for backup from fifty thousand brain researchers, and he presents their position fairly. In the field of neuroscience the mind is considered to be a by-product of the brain, the way sweat is a by-product of burning calories or flushed cheeks a by-product of sexual excitement. But thoughts are not readily broken down into data. Love and beauty aren't reducible to data, either. As the eminent British physicist Russell Stannard writes, "There is no way we can see concepts like *hope, fear,* and *pain* being quantified." In order to follow Christ's injunction to seek the kingdom of heaven within, or the Greek ideal to know thyself, the road lies only through the mind. And so spirituality puts mind first, where it belongs.

So how did the brain manage to dethrone the mind? Twenty or

thirty years ago the human brain was still poorly understood. One neurologist quipped that we knew so little about memory that the skull might as well be filled with sawdust. But the advent of new technologies shot brain research forward, and today a scan with an fMRI (functional magnetic resonance imaging) machine not only reveals the brain's memory centers; it can show them lighting up in real time, or going dark if a patient is suffering from Alzheimer's disease. Hope, pain, and fear may not be quantifiable, but at least we can film images of them as brain activity.

The logic that places brain before mind is amazingly weak, however. Let me give an analogy: I'm sure you would agree that you can't play "Twinkle, Twinkle, Little Star" on a piano without a piano. That's obvious, just as obvious as the fact that you can't have a thought without a brain. But if somebody told you that the piano *composed* "Twinkle, Twinkle, Little Star," the statement would make no sense. A piano is only a machine; it doesn't create new music. You can't overturn this fact by examining the molecules inside each ivory key under an electron microscope to explain where Mozart comes from, but brain researchers do just that when they probe the molecular structure of neurons for the hidden origin of thoughts and feelings. Before a piano can produce music, a mind must write the notes. Before a brain can register a thought, a mind must think it.

For centuries the mystery of how the mind relates to the body has been a philosophical question, not a practical one. So far as ordinary life goes, brain versus mind isn't a pressing debate. We say, "I've made up my mind," not "I've made up my brain." The average person goes through life never questioning that it takes a mind to be human. But this seemingly ivory-tower issue has incredibly practical implications. You cannot be indifferent to the question of mind versus brain if the mind serves as a portal to a deeper reality; if reaching that reality can transform your life, mind versus brain turns into the most urgent question of all.

We don't lack for inner voyagers. Neuroscience has already shown that the brain scans of advanced Buddhist monks are very different from the norm. (Earlier I mentioned the finding that the monks' brains operated at twice the frequency of normal brains in the gamma wave region.) The biggest discovery was that general activity in the prefrontal cortex was very intense—more intense, in fact, than ever observed before—a change that came about after years of meditation on compassion. As it happens, the prefrontal cortex is the brain's center for compassion, among other higher functions. In this case, it would be inaccurate to say that the brain changed itself. First the monks had the intention to be compassionate; they meditated upon it for years, and their brains followed suit.

This is the opposite of what science expected. One much-publicized view among Western doctors has been that visionaries like Saint Teresa of Avila and Saint Bernadette, figures who have had mystical experiences, might have suffered from brain lesions, epilepsy, or some other malady that fooled them into thinking they were experiencing God. (Among confirmed atheists, the way to explain a holy vision comes down to a choice between hoax, delusion, and brain disease. The last is actually the most compassionate explanation.) Skeptics can argue all they want about how an unbalanced brain fools mental patients into believing in illusions. Some schizophrenics with grandiose delusions believe that they can make a locomotive stop by standing in front of it and willing it to stop. Faith healers believe that they can cure cancer by asking God's help. Scoffers call such beliefs magical thinking. Everyone knows that you can't move objects with your mind. Yet that is precisely what you do when you make a fist or throw a ball: not only does your mind move thousands of molecules in the brain, but your intention spreads throughout the nervous system, reaching the muscles and bones—every step of the way is mind over matter. As for equating saintliness with mental disease, such a judgment is insulting and foolish on the very face of it.

What does matter is a strong desire to be close to God. As we saw with Tibetan monks, intention translates itself into new brain functioning. Why is that so incredible? Nobody can explain why we have *any* thoughts, so it's not that experiencing God is more mysterious than experiencing orange juice or the World Series. We can't shift in a spiritual direction unless the brain shifts, too, and it's our desire that alters the material landscape of the brain, not vice versa.

In many ways neurology is a red herring when deciding what a valid experience amounts to, since the visual cortex lights up when you actually see a horse and when you dream of one. An image is an image is an image, to paraphrase Gertrude Stein. Spirituality embraces a wider perspective. The cosmos didn't have to wait billions of years before the human brain evolved. The cosmos was behaving mindfully long before that. Here's eminent physicist Freeman Dyson: "It appears that mind, as manifested by the capacity to make choices, is to some extent inherent in every electron."

So which came first, mind or brain? Science is used to solving hard problems, but this one, as Leonard notes, is considered *the* hard problem. I'd like to propose that pitting mind versus brain is a no-win proposition. The hard problem can be settled without either side losing. Why must we claim that mind creates matter—or vice versa—in the first place? Such a need disappears once we concede that there is no entry point in the last 13.7 billion years when matter suddenly learned to think and feel. When we stop futilely searching for that fictitious moment, a better answer appears: mind has always been here, if not eternally, then as long as gravity and the laws of Nature have been here.

In this alternate view, the cosmic mind surrounds us so completely that no matter what we do it won't go away. It exists in our heart, liver, and gut cells as much as in our brain, providing intelligence, organizing power, creativity, and everything else. Even if you lose your mind through psychosis, drugs, or a catastrophic accident, the aspect of intelligence that keeps the body going will be intact (as we witness with patients in a coma). This neatly solves the chicken-or-the-egg riddle about which came first, mind or brain. "Coming first" isn't valid or relevant in the quantum vacuum, which is outside space and time. If gravity and mathematics began there, it's a small step to give mind the same status. After all, there's no way to experience mathematics, gravity, or anything else without a mind.

I realize that this small step carries science where many don't want it to go, into the realm of things that cannot be quantified. But science is already there. (A personal aside: I once discussed consciousness with a prominent physicist deeply versed in the hard problem. When I asked if he wanted to discuss the issue publicly, he shrank away. "You don't understand. Consciousness is the skeleton in the closet. We don't discuss it, and if I did, my professional reputation would be ruined.") Rumi, the beloved Sufi mystic, understood that mind is everywhere when he said, "The whole universe exists inside you. Ask all from yourself." Placing the mind center stage in the universe solves a vexing riddle wrapped inside the hard problem, as follows: When I see a sunset in my mind's eye, its glowing orange splashing across a sapphire sky, where is that sunset? It's not in my brain, because the brain has no light or pictures inside it. There is nothing in the brain but soft quivering tissue, pockets of water, and stygian darkness. Yet the sunset I envision has to be somewhere, and the best answer is mental space.

In mental space mind and matter move together as one. If I want to remember my mother's face, I conjure it up instantly. It doesn't matter how many thousands of neurons must be orchestrated, or what centers of the brain must light up, to turn memory into a visible image. Mind and matter are inseparable. As the instrument of consciousness, the human brain needed time to evolve. Once it evolved sufficiently, a thought and a neuron became connected as perfectly as a pianist and a piano—only in this case, the brain plays the music of life.

Leonard offered Maxell's Demon to defend the basic laws of physics. I have no difficulty with that, so long as "basic laws" include the quantum world from which all possibilities spring. Let me offer Deepak's Demon to defend the mind. This demon is perched on the top of the Empire State Building peering down at the traffic. Cars heading along Fifth Avenue sometimes turn left and sometimes turn right. The demon knows that all the cars are obeying the laws of physics, as are the atoms inside the bodies of the drivers. He knows that a statistical prediction can be made about which car might turn left or right. Does that mean that the laws of probability tell us what each driver is doing? Not at all, because Deepak's Demon realizes that each car represents a mind making a decision. Am I going to Macy's or to the United Nations? One is left, the other is right. Without the mind deciding first, cars don't turn.

So the hard problem can be solved, but it takes a broader vision to do it. Reductionism isn't enough. When asked what the quantum world means for everyday life, physics generally shrugs and goes about its day-to-day business. This attitude has been summarized as "Shut up and calculate." Physics is proud of its desire to remain aloof from metaphysics. But like it or not, we must bring the essence of existence center stage. Our minds cannot rest until we know what the mind is. Spirituality has always welcomed that quest; now it's time for science to do so as well.

13

Does the Brain Dictate Behavior?

DEEPAK

The average person can't easily be talked out of free will. If you go to a Chinese restaurant, you get to choose from column A or column B. You do not feel that someone or something is choosing for you. The universe runs according to physical laws, but we are still free to make our own choices. We may doubt our judgment afterward, it's true. Falling into a bad habit shows how some choices stick around and cannot easily be changed. Addictions go a step further. They make us feel that we are slaves to our craving and have no choice but to obey.

Spirituality is about widening your choices. Science can aid in this project or hold it back. It aids by giving us control over mechanical switches, whether they are in the brain or in our genes. It holds back the project when it insists that our brains or genes control us. No issue is more critical, because ultimately there is only one master, either you or the mechanisms built into your body. Most of us have not faced this issue head-on. We exert choice some of the time and run on automatic the rest of the time. Hence the resistance to posting the nutritional facts about a Big Mac on the menu. Nutrition involves thinking; fast food is mindless. Sometimes we are clear, sometimes confused, sometimes in charge, sometimes the victim of our conditioning. But life doesn't have to be so compromised.

At the moment, mainstream science is highly deterministic. As Leonard notes in an earlier essay, "Our choices are far more automatic and constrained than we'd like to think." I find that assessment both gloomy and unrealistic. In a brain scan, the same area of the prefrontal cortex associated with the motherly feeling of nurture lights up when the subject see photos of a baby or a puppy. A determinist would say that an identical reaction is taking place. But when you walk into a room where a baby is present, you don't give it a dog treat and burp the Irish setter. We override our brains all the time.

This is incredibly important, because it's all too easy to give up your power and lapse into unconsciousness. When you sit down with a bag of potato chips and eat the whole thing without noticing what you're doing, you have gone unconscious. When you let another person dominate or even abuse you because you don't want to make waves, you've gone unconscious, too. Reclaiming the power of choice comes down to reclaiming consciousness; the first step in this process is that you must want to be awake, aware, flexible, and free of old habits.

Neuroscience doesn't help in this regard when it reduces thinking and feeling to chemical reactions and electrical signals in the brain. Pathways that supposedly dictate behavior are mapped on MRIs and CAT scans. By now everyone has seen TV programs showing how a normal brain lights up compared to a distorted brain, the distortion being anything from a brain tumor, depression, or insomnia to criminality or schizophrenia. Such findings cannot be dismissed, of course. The mind has no choice but to be yoked to the brain, and when the brain is physically out of balance, mental changes will occur. But this is far from saying that the brain controls the mind.

Your behavior is constantly being influenced from many angles, both within and without. Indeed, one proof that the brain doesn't control the mind is that the brain lights up the same way when you remember a stress, like a bad auto accident or being fired from your job, as when you actually go through the stress. But we have no trouble knowing that a memory isn't the same as the real thing. Some determinists claim that thinking must be rooted in brain chemicals because the two are exactly correlated. A rush of adrenaline appears when a person becomes suddenly excited or afraid. The physical signs of fear are undoubtedly triggered by adrenaline, but that's not the same as saying that adrenaline, or any other chemical, causes fear.

Let's go into this a bit more deeply. There's a 2010 study from the Mount Sinai School of Medicine on the link between a hormone called oxytocin and how grown children feel about their mothers. Oxytocin, popularly known as "the hormone of love" because it appears in higher levels when people are in love, is found throughout the body; in the brain it has been associated with a number of positive things like trust, sexual pleasure, and low anxiety. When mothers give birth, oxytocin levels rise in the brain, which is connected to a powerful feeling of nurturing. Mothers who reject their babies or feel postpartum depression seem to lack this burst of oxytocin.

Chemical determinists would appear to have a powerful argument for their case, saying that oxytocin causes people to feel better in various ways, and their elevated mood leads to more positive thoughts. For example, a dose of oxytocin will make people act and feel more generous when they are put in a situation where they can choose to be generous or not. So does low oxytocin make a Scrooge and high oxytocin a philanthropist like Warren Buffett? That would indeed be deterministic. The new study casts serious doubts, however. When adults who had happy relationships with their mothers were given oxytocin, they remembered having even more positive feelings. But here's the rub. When subjects reported that they had a bad relationship with their mothers, a dose of oxytocin increased those bad feelings. The "hormone of love" has a dark side. More to the point, there isn't a one-to-one correlation to loving feelings, much less an established cause.

I've already mentioned the crudest metaphor used by proponents of artificial intelligence, that the human brain is a machine made of meat. Many brain researchers don't take this as a metaphor but as literal fact, to which there is a simple but devastating reply: a machine can't decide *not* to be a machine, but we do, all the time. Our nervous system can run the body on automatic pilot—that's why patients in a coma aren't dead—but if you're not in a coma, the same nervous system can release the controls over to the mind. To say that it's the machine itself that decides when to be in control and when not to be defies common sense; that's like having a car engine decide "It's my turn to drive."

The existence of free will, along with mind over matter, was once supported by neuroscience. In the 1930s a pioneering Canadian brain surgeon named Wilder Penfield discovered that if you stimulate the area of the brain that controls large muscles (the motor cortex), those muscles move involuntarily. In one trial, Penfield inserted a delicate wire into the specific area of the motor cortex that controls the arm, and when he sent a tiny shock through it, the patient's arm would shoot up. He would ask his patients what just happened. Their response was "My arm just moved." (Brain surgery is regularly performed with the patient awake and conscious, because the inner tissue of the brain feels no pain.)

So far, Penfield's results sound highly deterministic. He showed a causal link between brain and body, and it would seem to be only a small step to say that the brain must be controlling the body. But Penfield believed in the existence of the mind. He told his patients to raise their arms (without sending a small shock through the wire), which they easily did. Then he asked, "What happened now?" Their response was "I lifted my arm." In other words, patients knew the difference between "My arm just shot up" and "I lifted my arm." One is automatic, the other voluntary. It's deeply ironic, then, that brain researchers now defend the notion of determinism by replaying this same experiment to prove that the brain controls us when in fact it proves the opposite. (Penfield went on in his distinguished career to insist that the brain serves the mind.)

When people undertake spiritual disciplines like yoga, meditation, self-reflection, and devotion, they discover that it's possible to attain mastery even over involuntary processes. In a few minutes, for example, I could show you how to lower your metabolic rate and blood pressure through a simple exercise in focused attention. When carried to real mastery, meditation can slow heart and breathing rates almost to zero, a feat displayed by Eastern yogis and swamis. I could show you how to choose to make your palms warmer, or even to develop a red patch of hot skin on the back of your hand. Tibetan monks use their minds to warm their whole bodies sufficiently to sit all night in freezing Himalayan caves wearing only a thin silk robe. The worldview I'm arguing for wants people to move in the direction of such mastery.

What would you look like as a master? Let's ask the question with no religious overtones or exotic images of yogis and monks. Mastery means that you would be able to pursue self-determination—that is, you would have the freedom to write your own life script. There could be as many life scripts as there are people, but they'd all have one thing in common: a person's desires would increase his or her well-being. Right now, few of us can confidently match our desires with our well-being. We are severely limited by repetition and habit. This is where free will hits a wall, hard and often. But why?

You and I are spiritual paradoxes. Gifted with the most flexible nervous system in the universe, we tie it down with a thousand tiny ropes, just as Gulliver was tied to the beach by the Lilliputians. We are attached to our own little ways of doing things, our firm likes and dislikes, not to mention our memories, past conditioning, and emotional hot buttons that other people can push. A cognitive psychologist once calculated that 90 percent of the thoughts a person thinks today are the same as those he thought yesterday. We pay a high price for letting the nervous system run on automatic pilot.

It's tempting to blame our lack of mastery on the brain. Wedded to determinism, brain science used to make basic assertions that were eventually proven to be false. One such assertion held that the brain was inexorably hardwired for a given response. A good example is fear. When our ancestors were threatened by wild animals, they went into fight-or-flight mode, and the anatomical reason for that is our lower brain, inherited from ancestors as primitive as fish and reptiles. Stacked on top of the lower brain, exactly like an archaeological dig where new cities are stacked over the buried ruins of ancient cities, is the higher brain, or cortex. The higher brain is where we counter fear. We can look at a threat and tell ourselves, "Calm down. That wasn't a gun going off; it was a car backfiring," or "I'm scared, but I can't let my kids see it."

There are myriad ways for you to deal with fear through reason and higher emotions like devotion to family or a sense of duty. But fear comes first. Fight or flight has a privileged pathway in the brain, which is why you jump when a car backfires and think about it later. Thinking enables you to decide that the backfire was harmless. No need to fight or flee. In itself this two-part sequence seems beneficial. It's good to react quickly to danger, even if the danger turns out to be illusory. The problem is that if a reaction is repeated often enough, it forms fixed tracks in the brain, neural pathways that work automatically, curtailing freedom of choice. Each of us knows what it's like to lose control over our anger, eating habits, weight, anxiety, depression, and cravings of every kind. There's wisdom in the Talmudic saying, "No man owns his instincts." But civilization teaches us how to make them our allies, not our enemy.

In spiritual terms, losing control is traced to falling asleep. Strong materialists believe that the brain runs the show anyway: being awake (i.e., more free to choose) is a fairy tale we tell ourselves. They believe we are marionettes that refuse to see the strings that control them, and since the brain pulls invisible strings made of chemicals and electric signals, we are fooled into believing that our feelings of love, courage, and kindness, and our aspirations, have any force or meaning.

But what about the obvious fact that some people manage to break their old habits, overcome past conditioning, work through fears, and recover from addictions? Obeying a habit and kicking it are opposites. It cannot be true that the brain rigidly dictates behavior A and the opposite of behavior A. Inevitably brain science has had to soften its insistence on hard wiring, leading to a theory of soft wiring, which allows the brain to change the way a person wants it to change. The technical term for this is "neuroplasticity," which refers to how neural pathways can be altered at will.

Suddenly the prospect of mastery opens up enormously. A spectacular example involves going blind. Contrary to popular belief, blind people don't plunge into total darkness. Inner sight of some kind generally remains. One man blinded by a spray of industrial acid went on to envision and develop intricate gear boxes with dozens of interlocking parts. Another took up roofing and alarmed the neighbors by doing his work on extremely steep gables that he climbed at night. Sometimes another faculty takes over from sight. I once read about a blind marine biologist whose specialty was gathering highly poisonous sea snails in the Indian Ocean; he found the creatures with his toes, used touch to identify them, and never got poisoned. Eventually these inspiring examples of neuroplasticity led to a new technology, known as BrainPort, which gives the brain a controlled way to replace one sense with another.

The BrainPort device, which resembles a cap outfitted with electrodes, began as a wired-up chair mounted with a camera above and a pad on the back of the blind person that delivered a pattern of electrical signals to the skin. The person sitting in the chair would receive an image of what the camera saw by having the image sent to his back through the sense of touch. The brain transformed the "felt" image into a "seen" image. This breakthrough, which happened forty years ago, showed that one sense can substitute for another.

Later, neuroscientist Paul Bach-y-Rita, having made this breakthrough, found a way to restore balance to people whose brains had been damaged in that region. Losing your sense of balance is very disorienting, like walking perpetually on a rocking ship at sea. Bach-y-Rita placed a small pad on their tongue that sent a tiny electrical signal to the right, left, front, or back of the tongue, depending on which way the unbalanced person was tilting. His subjects quickly learned to bring the signal to the middle of the tongue, which meant that they were upright. After a while, the brain took over the task by itself. A person who previously couldn't stand up without falling over, now could be weaned off the BrainPort and walk, or even ride a bicycle, on their own.

The brain is guided by determination, as was learned very early by the Bach-y-Rita family. In 1959 Paul's father, Pedro, suffered a debilitating stroke that paralyzed one side of his body and impaired his speech. A second son, George, was a psychiatrist, and by defying the prevailing belief back then that such damage would be irreversible (the brain wasn't supposed to be able to heal itself), George helped his father regain a normal life. Years later, when Pedro died, his brain was examined, and it was found that the brain stem, the portion damaged by the stroke, had in fact repaired itself.

One aspect of science can be thanked for these discoveries, even if another aspect clings to determinism. The fork in the road could hardly be clearer. If you and I choose to attain mastery, our spiritual goal finds a physical ally. The human brain, like the universe itself, delivers whatever you expect it to, in accordance with your deepest beliefs. So why not believe that your brain can deliver mastery? If one sense can be substituted for another, if the brain can heal itself, and if new neural pathways develop because a person decides they can, there is much more freedom available to us than anyone ever supposed.

LEONARD

In his book *The Incoherence of the Philosophers,* the eleventh-century Sufi philosopher Abu Hamid al-Ghazali wrote that when fire is held to cotton, the cotton is not burned by the fire, but is burned directly by God. According to this view, our expectation that the fire causes the cotton to burn arises because each time that we have placed cotton in fire, God has willed that the cotton burn; but the fire itself cannot dictate the burning because that would tie God's hands, and God is free to do whatever God wishes. More generally, al-Ghazali argued that the laws of nature are a kind of illusion we've come to believe in because God is rational and usually consistent (except in the cases of miracles). The connection between cause and effect only *seems* to follow unalterable laws, with the true causes of events lying beyond our physical realm.

Deepak and many others have an attitude very similar to this when it comes to the connection between the physical brain and human consciousness. We can study the brain and understand its laws, but in their view the physical substrate of our cortex is ultimately controlled by an invisible hand of consciousness that is the true source of our thoughts, feelings, and actions. Deepak believes that the brain is the puppet of the immaterial mind—which because it is immaterial is not governed by physical laws.

Deepak compares the neurons in our brain to a piano, and our conscious mind to the music it plays. In this view, consciousness is expressed by our physical brain just as musical notes are brought to life by a physical piano. Deepak says, "You can't play 'Twinkle, Twinkle, Little Star' on a piano without a piano. . . . But if somebody told you that the piano *composed* 'Twinkle, Twinkle, Little Star,' the statement would make no sense." That is true. But if somebody told you that "Twinkle, Twinkle, Little Star" was composed in an immaterial realm of universal consciousness, that would also sound illogical—and *that,* if you follow Deepak's analogy, is the alternative he offers.

Let's not be misled by analogies. While both viewpoints—that consciousness comes from an outside realm, and that it arises within the brain itself—are admittedly challenging, the way to make progress in elucidating the connection between mind and brain is to *examine* the brain, and see how much of what we do and feel can be accounted for by its actions. Deepak writes that you can't understand anything about the connection between a piano and how the music that is played on it is made by "examining the molecules inside each ivory key under an electron microscope," which he believes is comparable to what brain researchers are trying to do when they look at the brain in order to find a physical basis for the mind. But when one looks at the brain, one *does* find that there is plenty of evidence to indicate that the brain is the source of consciousness.

Deepak and I have been doing all the work so far, which is only fair, since we are the authors. But here is a little exercise for you. Have a look at the blocks pictured on the next page. One of the black tops looks long and narrow, the other shorter and wider. They aren't—if you measure them, you'll find they are identical. You are fooled because the perspectives in the drawings were designed to take advantage of a quirk in the way your brain perceives shapes. Now please look at the blocks again, and, now that you know they are identical, try to see them that way. You'll find that you can't. Such illusions and the inability to overcome them are evidence that there is no external mind separate from the physical brain and capable of lording over it. We cannot transcend the workings of the physical brain.

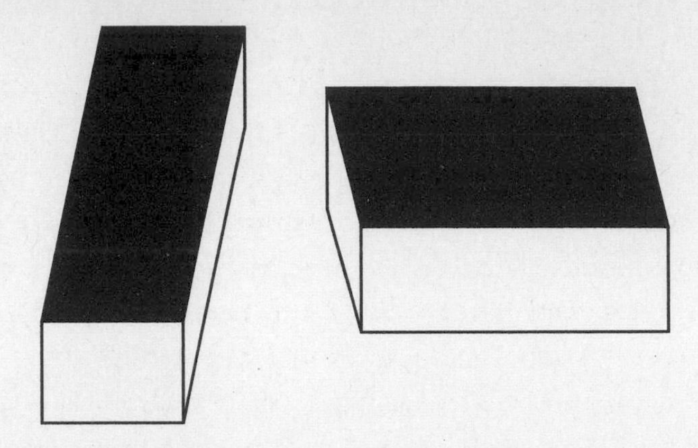

Here's another example. Have a look at the faces below. How do they strike you? A male and a female of roughly equal attractiveness, with the woman on the right? We all have our own quirks when it comes to making assessments of attractiveness, but the first requirement for a successful love life is being able to recognize the sex of your preference when you see it. And if you think the faces below belong to people of different sexes, you are wrong. They are the same face, differing only in the degree of contrast in the photograph. In both East Asians and Caucasians—the populations studied—female faces exhibit greater contrast, and though that is probably news to you, it is not news to your brain. It automatically interprets the image with less contrast as male, and even after you know the faces are identical, it is difficult or impossible to override your brain's automatic judgment.

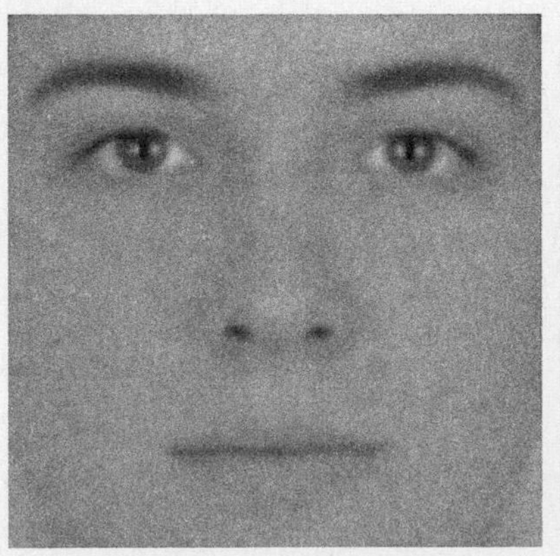

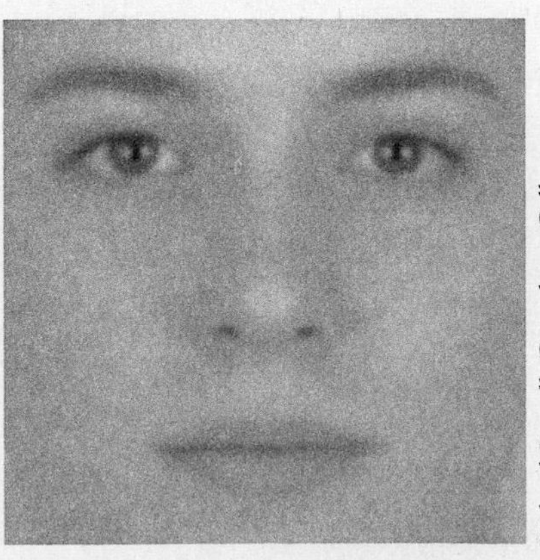

Richard Russell, Gettysburg College

There are also many striking examples of the deterministic connection between the brain and the mind in nonhuman animals. I mentioned the courtship of fruit flies last chapter. It accounts for most of their social life and is, in one researcher's words, "the activity they do best." The male's usual courtship behavior is to approach the female, tap her with his forelegs, vibrate his wings, lick her, and then curl his abdomen and wait. If she is interested, she'll approach, and if not, she'll buzz her wings at him. How can one account for the swagger of the fruit fly? It has been traced to a gene that causes a particular protein to be created in certain neurons within the fly's brain. Those neurons appear to direct each step of the coordinated sequence of courtship. For example, when a biologist genetically engineered female flies to make the male version of that protein, the females aggressively pursued other females, and performed the male courtship dance.

Mammals, too, can be manipulated chemically or genetically in a manner that seems to reduce them to robots. For example, though female sheep—ewes—can be downright nasty to strange lambs, they appear to be caring and loving mothers to their own babies. As it turns out, their admirable maternal behavior is directly traceable to the oxytocin that is released in the mother's brain when it gives birth. During the period in which her oxytocin level is elevated (which after giving birth lasts about two hours) a ewe will suckle and bond with any lamb that approaches, learn its smell, and then proceed to raise it to adulthood whether or not it is her own. Outside that window of time, however, a ewe will chase off any lamb she has not previously bonded with—even her own infant if the new baby was withheld until her oxytocin levels fell. Moreover, the ewe's bonding behavior can be turned back on at any time through an injection of oxytocin.

Other animals in which the role of oxytocin has been studied extensively are the voles, a group of about 150 species that resemble mice. One type of vole, the prairie vole, is a loyal mate that forms bonds for life and rarely takes on a new partner, even if its original partner disappears. Two other species of vole, however, the montane and meadow voles, are promiscuous loners. As in sheep, the behavior of these animals can be traced to oxytocin, and to a related compound called vasopressin. Increasing the level of these chemicals in the brain of a promiscuous montane or meadow vole will make it a model husband and father, while decreasing the level in the prairie vole will cause it to act more like its loner cousins. Interestingly, scientists have found a gene that governs vasopressin receptors in the human brain and have observed that it causes differences among humans analogous to the differences between the voles. Men who fall into the montane/meadow vole category in terms of their vasopressin levels were found to be twice as likely to have experienced marital problems, and half as likely to be married.

Deepak asks, "So does low oxytocin make a Scrooge and high oxytocin a philanthropist like Warren Buffett? That would indeed be deterministic." We can obviously perform only limited experiments on humans, but when oxytocin brain levels are manipulated in animals, the answer has been yes, such manipulations *do* result in the corresponding behavioral changes.

The relationship between oxytocin and behavior in people is of course far more complex than in these animals. As Deepak mentions, in humans oxytocin also seems to have a connection with certain negative feelings. That is not a sign that brains don't determine behavior. It means only that brains are complicated, and hormones play many roles. But in human mothers, as in ewes, oxytocin is released during labor and delivery, and promotes bonding.

That the brain directs behavior and emotions is also sadly evident in people whose brains have been damaged. Nowhere is the effect of an altered brain on behavior starker than when it impacts a person's moral judgment. "Moral judgment is, for many, the quintessential operation of the mind beyond the body, the Earthly signature of the soul," wrote neuroscientist Joshua Greene. But Greene and other scientists have made a lot of progress in understanding how the physical brain creates moral judgment just as it encodes memories or interprets visual information. One area of the brain vital to that function is called the ventromedial prefrontal cortex, or VMPC, which sits just inches behind the forehead. Patients with severe impairment of the VMPC have unchanged intellectual abilities, but they exhibit less empathy and a reduced revulsion to hurting others. In one study a group with VMPC injuries and a control group were presented with a series of hypothetical moral choices involving the killing of an innocent person for the greater good. Those with VMPC injuries were twice as likely to say they'd push someone in front of a train to save a group of others or suffocate a baby whose crying threatened to attract enemy soldiers. In real life VMPC damage has been associated with the onset of divorce, job loss, and inappropriate social conduct. In fact, many habitual criminals are psychopaths who typically begin to exhibit cruelty in their early years, and display shallow emotion and lack of empathy throughout their lives. Neuroscientists have found a neural basis for their behaviors, implicating a wide range of brain regions such as the VMPC and the amygdala. "Because of their brain damage, these patients have abnormal social emotions," said neuroscientist Ralph Adolphs, one of the VMPC researchers.

We commonly accept that physical disability in stroke victims is due to brain damage, but the prospect of viewing "evil" as a neurological deficit, the direct result of a person's brain structure, can be unsettling. It may feel as though we are excusing the individual ("his brain made him do it"). There is one group, however, for whom we do readily make allowances for moral or ethical lapses traceable to insufficient development of the prefrontal cortex. This is an easily identifiable group, and one close to the heart of many of us. I'm referring, of course, to children. We recognize that below a certain age children should not be treated as responsible adults or held accountable in the same way. Our legal system makes this distinction, and so do most of us—the main reason for this being that the prefrontal cortex isn't fully developed until the early twenties. The risk-taking behaviors of teenagers, and their lack of impulse control in the face of the urge for immediate gratification, are common knowledge, and now we know not just that they exist, but why.

I agree with Deepak that human behavior "is constantly being influenced from many angles." Those angles include past experiences and current circumstances, and their influence on the many brain structures whose complex interactions create the people we are. But all those angles are within our physical world. There is no evidence that, as Deepak believes, our brains are controlled by something outside them. Still, we are not slaves to our genes. People can change, and I agree with Deepak that "when people undertake spiritual disciplines like yoga, meditation, self-reflection, and devotion, they discover that it's possible to attain mastery even over involuntary processes."

Neuroscience doesn't debunk those ideas; it provides support for them. In fact, the studies of Buddhist monks that have shown how they can modulate the activity of their brains are illustrations of a feedback loop. Like the experimental subjects I mentioned in chapter 12 who could make their neurons fire at will in order to control images on a computer screen, the monks offer another example of a decision of the mind-brain system that can alter the brain.

Mastery, self-determination, and the freedom to write our own life script are admirable goals, and I believe that we—that is, our brains—can achieve those goals. And they don't have to leave the material world to do it.

14

Is the Brain Like a Computer?

LEONARD

In 1955 a group of computer scientists appealed to the Rockefeller Foundation to fund a meeting of ten experts at Dartmouth College. The scientists said they intended to "proceed on the basis of the conjecture that every aspect of learning or any other feature of intelligence can in principle be so precisely described that a machine can be made to simulate it. An attempt will be made to find how to make machines use language, form abstractions and concepts, solve kinds of problems now reserved for humans, and improve themselves." They stated their agenda clearly and concisely, but the most striking sentence of their proposal is the one that followed their agenda statement. They said, "We think that a significant advance can be made in one or more of these problems if a carefully selected group of scientists work on it together for a summer." In retrospect, it seems obvious that significant progress in artificial intelligence comes over decades, not "a summer." As cognitive neuroscientist Michael Gazzaniga put it, they were "a little optimistic."

At the very heart of this early overoptimism is the "brain-as-computer" metaphor, which is, at best, an oversimplification. The operating characteristics of biological brains are very unlike those of the computers that were used in 1955, or even of the much more sophisticated ones we build today. Conventional computers consist of electronic components such as transistors—a kind of on-off switch—that implement a series of logical operations called gates. Logician George Boole proved in 1854 that any "logical expression," including complicated mathematical calculations, can be implemented by a "logic circuit" made by wiring together components built from just four fundamental gates, called AND, OR, NOT, and COPY. These gates transform one or two bits of information at a time (a bit is a register—a storage location—that can have the value 0 or 1). For example, a NOT gate changes a 0 to a 1 and vice versa, while a COPY gate changes 0 to 00, and 1 to 11. Whatever it is used for, all a computer is really doing is applying electronic logic gates to bits, one or two at a time. Brains, on the other hand, execute operations in a parallel manner, doing millions of things simultaneously.

There are many other distinctions. Brains' processes are noisy—that is, subject to undesired electrical disturbances that degrade useful information—while computers are reliable. Brains can survive the removal of individual neurons, while a computer operation will fail if even a single transistor it employs is destroyed. Brains adjust themselves to the tasks at hand, while computers are designed and programmed for each finite task they must perform. The physical architectures, too, are quite different. The human brain contains a thousand trillion synapses, while a multimillion-dollar system of computer hardware today might have a trillion transistors. Moreover, though synapses (the gaps between neurons through which electrical and chemical signals flow) are a bit like transistors, a neuron's behavior is vastly more complex than that of a computer component. For example, a neuron fires—sending its own signal to thousands of others—when the aggregate signals from the neurons that feed it reach a critical threshold, but the timing of the incoming signals matters. There are also inhibitory signals, and neurons can contain elements that modify the effect of incoming messages. It's an intricate design of vastly greater richness and complexity than anything employed in electronic devices.

Still, a metaphor can be useful even if the things being compared correspond in just one aspect. Carson McCullers wrote that "the heart is a lonely hunter," and that is a wonderful observation despite the fact that hearts don't carry rifles. So it can be helpful to think of the brain as a computer despite the differences in physical design and operation if, for example, biological brains and computer "brains" produce similar behavior. Among simple animals and advanced (by today's standards) computers that can certainly be the case. Take the female hunting wasp, *Sphex flavipennis.* When a female of that species is ready to lay her eggs, she digs a hole and hunts down a cricket. The expectant mother stings her prey three times, then drags the paralyzed insect to the edge of the burrow and carefully positions it so that its antennae just touch the opening. After the cricket is in place the wasp enters the tunnel to inspect it. If all is well, she drags the cricket inside and lays her eggs nearby so that the cricket can serve as food once the grubs emerge. The wasp's role as mother completed, she seals the exit and flies away. Like the ewes I described in the previous chapter, these female wasps appear to be acting thoughtfully, and with logic and intelligence. But as the French naturalist Jean-Henri Fabre noted in 1915, if the cricket is moved even slightly while the wasp is inside inspecting the burrow, when the wasp emerges she will reposition the cricket at the entrance, and again climb down into the burrow and look around—as if she had arrived with the cricket for the very first time. In fact, no matter how many times the cricket is moved, the wasp will repeat her entire ritual. It seems that the wasp is not intelligent and thoughtful after all, but rather follows a hardwired algorithm, a fixed set of rules. Fabre wrote, "This insect, which astounds us, which terrifies us with its extraordinary intelligence, surprises us, in the next moment, with its stupidity, when confronted with some simple fact that happens to lie outside its ordinary practice." Cognitive scientist Douglas Hofstadter calls this behavior "sphexishness."

If living creatures can appear intelligent, but disappoint when they sink to the level of sphexishness, digital computers can excite us when they *rise* to merit that same modest label. For example, in 1997 a chess-playing machine named Deep Blue beat reigning world chess champion Garry Kasparov in a six-game match. Afterward Kasparov said he saw intelligence and creativity in some of the computer's moves and accused Deep Blue of obtaining advice from human experts. In the limited domain of chess, Deep Blue seemed not only human, but superhuman. But although the human character Deep Blue displayed on the chessboard was far more complex, nuanced, and convincing than the motherly care displayed by the wasp, it did not arise from a process most of us would be likely to think of as intelligent. The three-thousand-pound machine made its humanlike decisions by examining 200 million chess positions each second, which typically allowed it to look six to eight moves ahead, and in some cases, twenty or more. In addition, it stored a library of moves and responses applicable to the early part of the game, and another library of special rule-based strategies for the endgame. Kasparov, on the other hand, said he could analyze just a few positions each second, and he relied more on human intuition than on processor power. Even without checking under the hood, there is an easy way to illuminate the differences in intelligence: just change the game a bit. For example, scramble the pieces' starting positions—or eliminate the rule, important in the endgame, that allows a pawn to be traded for any more powerful piece if it advances to the opposite end of the board. Kasparov would be able to adjust his thinking accordingly. But Deep Blue would be more like the wasp, unable to adapt to circumstances and make a judgment, its enormous apparent intelligence suddenly decimated by its inflexibility.

Deep Blue had a superhuman ability in chess, but it wasn't what most of us would term "intelligent." The same can be said of Watson, IBM's *Jeopardy*-playing computer that in 2011 beat the best human champions. To equip it for the game, IBM stuffed Watson with 200 million pages of content stored on 4,000 gigabytes of disk space, and endowed it with 16,000 gigabytes of RAM and an estimated 6 million rules of logic to help it arrive at its answers. Still, though Watson was usually right, it got to the answers through brute-force searches based on statistical correlations, not on anything one could call an "understanding" of the question. That issue was best illuminated by some of its *wrong* answers, such as choosing Toronto as an answer in the category "U.S. Cities." Behind its vast fact base, and impressive ability to answer questions posed in plain colloquial language, there was actually nothing one could call real intelligence.

Jeopardy is a relatively recent invention. So, too, are computers. That a computer could be programmed to play chess like an intelligent human being was originally proposed by mathematician Alan Turing in 1941, before the first machine that could reasonably be called an electronic computer had even been built. One of the most influential thinkers of the twentieth century, Turing provided many of the ideas that form the foundation of computer science, and of our digital age. He recognized the limitation of a computer having sphexish intelligence confined to a particular domain, and argued that thinking machines should also be learning machines, capable of altering their own configurations.

How does one judge whether a computer is intelligent? In Turing's view, intelligence should not be assessed according to the process through which beings or machines operate, but rather from the outside, by examining their behavior. That is how we judge people in our everyday lives—after all, we normally have no access to the inner feelings and thought processes of people we don't know. Little was known in Turing's day about the detailed workings of biological brains, but Turing believed that whatever the physical similarities and differences, a computer with intelligence and behavior similar to those of a human being would eventually be built.

Turing's life and brilliant career were cut tragically short when he committed suicide in 1954 after being forced by the British government to undergo hormone therapy because he was gay. But a few years before his death Turing published a paper titled "Computing Machinery and Intelligence," in which he outlined a test for intelligence that is now called the Turing test. In its current form, the Turing test consists of a human being and a computer that both interact through textual messages with an impartial human judge. If the judge cannot distinguish which contestant is the human, the machine is called intelligent.

Turing predicted that computers would pass this test within a half century, that is, before the year 2000.

The Turing test burst into popular culture in 1966, with a simple two-hundred-line computer program called ELIZA, which employed a handful of tricks to generate bland responses that often sounded startlingly humanlike. In the decades since, computer scientists have created many other "chatting robots." Here is a snippet of dialogue from a chat at a recent Turing test contest. Can you distinguish which speaker is the human judge and which is the robot contestant?

BILL: Do you know any good jokes?
JOAN: I don't even know any bad ones. I just can't remember them for very long.
BILL: Please tell a joke.
JOAN: What do you get if you walk under a cow?
BILL: I don't know.
JOAN: A pat on the head.
BILL: That's very good.
JOAN: I know.

The actual conversation went on for twenty-five minutes. The truth: Joan is the computer. But unlike the designer of ELIZA, which merely followed fixed scripts, Joan's creator took Turing's advice and designed this machine to achieve its "intelligence" through learning: the program "chatted" online over a period of years with thousands of real people, building a database of several million utterances which it searches statistically when composing its replies.

Computer scientists still haven't succeeded in creating a program that can consistently fool human judges over an extended period of time. But knowing both the degree to which programs like Joan do work, and *how* they work, suggests two conclusions. First, achieving "intelligence" of the Turing test variety in a digital computer is far more difficult than most people initially thought. Second, there is something wrong with the Turing test—for a machine that cobbles together speech by repeating responses it encountered previously isn't exhibiting intelligence any more than a nematode that slithers past a McDonald's is demonstrating culinary sophistication.

Though the Turing test is questionable, and has fallen out of favor with researchers in artificial intelligence, no better litmus test for intelligent thought has gained general acceptance. There are some interesting ones out there, however. Christof Koch and his colleague Giulio Tononi argue that—contrary to Turing's belief—the key point *is* to assess the *process* the being or machine in question utilizes, something easier said than done if you have no access to the candidate's inner workings. They propose that an entity should be considered intelligent if, when presented with any random scene, it can extract the gist of the image, describe the objects in it and their relationships—both spatial and causal—and make reasonable extrapolations and speculations that go beyond what is pictured. The idea is that any camera can record an image, but only an intelligent being can interpret what it sees, reason about it, and successfully analyze novel situations. To pass the Koch-Tononi test a computer would have to integrate information from many domains, create associations, and employ logic.

For example, look at the image on the facing page from the film *Repo Man*. An insect crawling over the page might detect the photo's purely physical qualities—a rectangular array of pixels, each of which is colored in some shade of gray. But in just an instant, and without apparent effort, your mind realizes that the picture depicts a scene, identifies the visual elements, determines which are important, and invents a probable story regarding what is transpiring. To meet the criteria of the Koch-Tononi test an intelligent machine ought to be able to key in on the man with the gun, the victim with raised arms, and the bottles on the shelves. And it ought to be able to conclude that the photo depicts a liquor store robbery, that the robber is probably on edge, that the victim is terrified, and that a getaway car might be waiting outside. (The scenes depicted would obviously have to be tailored to the cultural knowledge base of the person or computer being tested.) So far no computer can come close. An unintelligent brute-force approach like that which achieved limited success in passing the standard Turing test is of no help in passing the Koch-Tononi test. Even limited success in passing their own test, these researchers believe, is many years away. In fact, it was only a few years ago that computers gained the ability to do what a three-year-old child can do—distinguish a cat from a dog.

Is the fact that computers have had so little success thus far at

Edge City/Universal/The Kobal Collection

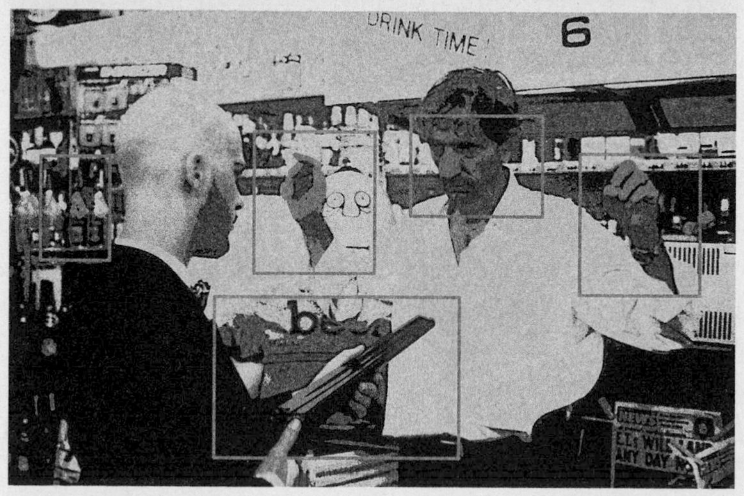

IEEE Spectrum

achieving the same sort of intelligence as our brain a technical problem, which we may one day solve? Or is the human brain inherently impossible to replicate?

In the abstract sense, the purpose of both brains and computers is to process information, that is, data and relations among data. Information is independent of the form that carries it. For example, suppose you study a scene, then photograph it and scan the photo into your computer. Neither your memory nor the computer's will contain a literal image of the scene. Instead, through an arrangement of their own physical constituents, mind and computer will each symbolize the information defined by the scene in its own trademark fashion. The information in the physical scene would now be represented in three forms: the photographic image, its representation in your brain, and its representation in the computer. Ignoring distortions and issues of limited resolution, these three representations would all contain the same information.

Turing and others turned such insights about information, and how it is processed, into an idea called the "computational theory of mind." In this theory, mental states such as your memory of the photograph, and more generally your knowledge and even your desires, are called computational states. These are represented in the brain by physical states of neurons, just as data and programs are symbolized as states in the chips inside a computer. And just as a computer follows its programs to process input data and produce output, thinking is an operation that processes computational states and produces new ones. It is in this abstract sense that your mind is like a computer. But Turing also took the idea a big step further. He designed a hypothetical machine, now called a Turing machine, that in theory could simulate the logic of any computer algorithm. That shows that, to the extent that the human brain follows some set of specified rules, a machine can indeed—in principle—be built that would simulate it.

The computational theory of mind has proved useful as a framework scientists can use to think about the brain, and technical terms common in information theory are now used widely in neuroscience, terms such as "signal processing," "representations," and "codes." It helps us to think about mental processes in a theoretical way, and to better understand how beliefs and desires need not reside in some other realm, but can be embodied within the physical universe.

Still, biological brains are not Turing machines. The human brain can do far more than simply apply a set of algorithms to data and produce output. As described earlier, it can alter its own programming, and react to a changing environment—not just to sensory input from the outside, but even to its own physical state. And it has astonishing resilience. If the corpus callosum is cut, severing the brain in two, a person doesn't die, but somehow goes on functioning, a wondrous testament to just how different we are from the computing machines that we build. A human brain can suffer the degradation of disease, or have vast sections obliterated through stroke or accidental impact, yet reorganize itself and go on. The brain can also react psychologically, and it is as resilient in its spirit as in its ability to heal itself. In *Stumbling on Happiness,* psychologist Daniel Gilbert wrote about an athlete who, after several years of grueling chemotherapy, felt joyful and said, "I wouldn't change anything," and about a musician who became disabled, but later said, "If I had it to do all over again, I would want it to happen the same way." How can they say things like that? Whatever happens, we find our way. As Gilbert says, resilience is all around us. It is just these qualities of the human mind that elevate it above simple algorithmic machines, providing both the beauty of being human and the greatest mystery that science has yet to unravel.

DEEPAK

The last time someone asked you if it looked like rain, did you reply, "I'll have to sample some randomized variables for that"? If a person came to you to translate the *Kalevala,* the Finnish national epic, would you say, "I'm sorry, that's not programmed in my software"? On the face of it, people don't think like computers, which are machines that shuffle two numbers, 0 and 1, to arrive at their "thoughts." Even if you believe, as Leonard apparently does, that the brain will eventually reveal the secrets of the mind, the brain doesn't operate using 0s and 1s, either. There is really no similarity between our brains and any "thinking" machine yet devised, which means that those quotation marks aren't going away.

Inevitably, the once promising field of artificial intelligence (AI) has not come close to reproducing actual thought. Leonard has covered the basic problems with AI, so I could just nod my head in agreement and move on. But there's a crucial question left hanging in the air. If the brain isn't like a computer, what *does* it do to produce thoughts? I believe the answer is clear-cut: the brain doesn't produce thoughts. It transmits them from the mind. What does the mind do, then? It creates meaning. Not only that, but meaning evolves, and as it does, the brain races to catch up, guided by the next interesting thing the mind wants to think about.

If a computer could embrace meaning, AI would make an earth-shaking breakthrough. Science fiction would become reality, since one of the favorite plots in science fiction consists of computers who outsmart their human masters, either turning on them or becoming all too human themselves. HAL the onboard computer stole the movie *2001: A Space Odyssey* by sounding more sympathetic than the robotic astronauts traveling into deep space. The audience was shocked when HAL decided to kill off the crew for the sake of the mission, and yet it was also touching when the last surviving spaceman started to dismantle HAL's memory, and the dying computer voice pleaded, "Please don't do that, Dave. I feel strange." Isaac Asimov's *I, Robot* explores the same theme, when mankind's mechanical slaves rebel against their masters.

The ability of computers to imitate us isn't just entertaining. One of the more ingenious software programs was ELIZA, already referred to by Leonard. ELIZA used a clever trick, based on a school of psychotherapy developed by psychologist Carl Rogers in the 1940s and 1950s, which put patients at ease by making empathic remarks of a seemingly simple kind, such as "I understand," "Tell me more about that," or just "Um." Programming such statements into ELIZA bypassed the computer's need to know anything about the real world. Bland, empathic remarks have the effect of making people feel heard and understood. Presto, a computer comes off as human. (In fact, various people who talked to their computers through ELIZA reported therapeutic results as good as those of a real psychiatrist.)

My position is that computers will never think—tricks can offer a good imitation, but no machine is capable of creating meaning, of crossing the line that separates mind from matter. However, the instant I make such a claim, a huge obstacle stands in the way. The brain is matter, and it seems to traffic in meaning. If squishy bits of watery floating chemicals in a brain cell can transmit the words "I love you" and await with exquisite vulnerability to hear if the other person will reply "I love you, too," a computer in the future may be able to do the same. Why not?

Rather than jumping headfirst into a complex argument about mind and meaning, let's consider the following experiment. Subjects at Harvard volunteered for a study in game strategy. They were seated in front of a monitor and told the rules of a specific game. "You are playing with a partner who is hidden behind a screen. There are two buttons each of you can push, marked 0 and 1. If you both press 1, you get a dollar, and so does your partner. If you both press 0, you get nothing, and so does your partner. But if you press 0 while your partner presses 1, you get five dollars, and he gets nothing. The game lasts half an hour. Begin."

Imagine yourself as a player of the game—what would your strategy be? Would you cooperate by pressing 1 all the time, so that you and your partner got the same reward? Or would you sneak in with a 0 while he was innocently pressing 1, so that you got a much bigger reward? You'd be tempted, but if he got angry enough, he could retaliate by pressing 0 all the time, forcing you to do the same, and then both of you would wind up with nothing.

After the experiment was conducted, subjects were asked about how their hidden partners played the game, and many said that their partners were irrational. Even when the subjects pressed 1 many times in a row, for example, signaling a willingness to cooperate, their partners refused. They would sneak in with a 0 in order to grab five dollars, while other times they seemed intent on pointless sabotage. It became necessary to punish them by pressing 0 all the time, but that didn't faze them, either.

In reality, this wasn't an experiment about game strategy at all. It was an experiment in psychological projection, because there were no hidden partners. Each subject played against a random number generator, which spewed out 0s and 1s in no particular order. Yet when asked what their partners were like, subjects projected human traits onto them, using words like "devious," "uncooperative," "fickle," "underhanded," "stupid," and so on. The human mind, it seems, creates meaning even when none is present.

The mind is all about meaning, and machines cannot travel there. Unless you have Beethoven on hand to input a Tenth Symphony, Shakespeare to input his lost play *Cardenio,* or Picasso to input a style of painting he never expressed on canvas, the machine is helpless to do so. Creative inspiration can't be reduced to writing code. Artificial intelligence was doomed from the start because "intelligence" was defined as logic and rationality, as if the other aspects of human thought—emotions, preferences, habits, conditioning, doubt, originality, nonsense, etc.—were beside the point. In fact, they are the glories of our highly fanciful, perversely delightful intelligence. Meaning has flowered through us in all its facets, not just as reason. These include irrationality. Atomic war is an example of such irrational behavior that it makes us shrink in terror from our own nature, but the *Mona Lisa* and *Alice in Wonderland* are just as irrational, and we gravitate toward them in fascination.

Computers are bound by rules and precedents, without which logic machines cannot operate. Computers don't say, "When I was daydreaming, something suddenly occurred to me." Yet Einstein did a lot of daydreaming, and the structure of benzene was revealed to the chemist Friedrich August Kekulé in a dream. (Somewhat ironically for AI, the German physiologist Otto Loewi, who won the Nobel Prize in Medicine in 1936, discovered how nerves transmit signals thanks to a dream he had.) So be grateful for the irrational. The French philosopher Pascal was right when he said, "The heart has reasons that reason cannot know."

I imagine that Leonard would agree with most of this. But I also imagine he would cling to the belief that one day a deeper understanding of the brain—he points in the direction of neural networks—will tell us what thinking is. Yet, what if no such solution exists? There may be no simpler model of the brain than the brain itself. This doesn't mean that the mind-brain connection isn't evolving. Certainly it is. When the mind created reading and writing several thousand yeats ago, a region of the cerebral cortex adapted and made reading and writing physically possible. When new forms of modern art are created, people scratch their heads at first, just as they did when Einstein's theory of relativity appeared, but in time they catch on, and then for future generations

Cubism and relativity become second nature, just as reading and writing are. Once you train your brain to read and write, you cannot go backward and reclaim illiteracy. Those black marks of ink on the page will forever be letters, not random specks. Irrevocably, meaning has moved you forward.

The spiritual life is entirely about moving meaning forward, and I contend that science alone will never be equal to that project. The fact that mind isn't matter goes to the heart of my argument, but so does a more technical point, which revolves around a famous mathematical argument known as Gödel's incompleteness theorems. In order to grasp what those theorems mean in everyday life, we must look into the nature of logical systems. We are the only creatures that love all kinds of nonsense. "'Twas brillig, and the slithy toves . . ." but sense is where we make our home.

In our craving for meaning, logic is our primary tool for determining what makes sense and what doesn't. But how can we truly know if we're right? The laws of nature make sense because they can be reduced to mathematics, a completely logical system. That's why we tell each other that two plus two equals four, not three or five. But can logic somehow fool itself? If so, then the world may seem to make sense when in reality it doesn't. (Thousands of years ago, the ancient Greeks were wrestling with this issue and ran into baffling riddles like the following paradox: A philosopher from Crete named Epimenides declares, "All Cretans are liars." Should you believe him? There's no way to know. He could be telling the truth, but that means that he's lying. Self-contradiction is built into the sentence.)

In simplified form, this is the problem that confronted Kurt Gödel (1906–1978), an Austrian mathematician who joined the wave of illustrious immigrants who escaped war-torn Europe to live in the United States. Gödel's area was the logic that governs numbers. We don't have to delve into that specialized field, except to say that natural numbers (the counting numbers like 1, 2, 3, etc.) are considered facts of nature and therefore can stand in for other things we take as facts. Numbers need to be consistent; when you apply procedures to them, the results should be provable. The same can be said of facts about the body, such as heart rate and blood pressure, because they, too, are governed by numbers. The doctor learns what range of numbers is considered normal, and your health is measured against that standard.

Gödel distilled numbers down to their purest essence, the logical processes that lead to such things as computers. What Gödel found is that logical systems have built-in flaws. They contain statements that cannot be proven—hence, his notion of incompleteness. His first theorem says that incompleteness is the fate of any logical system; there will never be a system that explains everything. His second theorem says that if you are looking at a system from the inside, it might be a consistent system, but you won't be able to find out as long as you stay inside the system. A blind spot is built in, because certain unprovable assumptions are part of every system. If you want to escape these fatal flaws, you must find a way to step outside the system. Logic cannot transcend itself.

Spirituality argues that consciousness can go where logic can't. There is a transcendent reality, and to reach it, you must experience it. Leonard, who is mathematically sophisticated, may be able to demonstrate how I've misconstrued these highly technical matters. But it's hard to escape one of Gödel's main points, that mathematical systems include certain statements that are accepted as true but which cannot be proven. If I boldly take this out of the realm of numbers, Gödel is saying that unprovable things are woven into our explanation of reality. Religionists make statements based on the assumption that God exists, although they can't prove it. Materialists make statements based on the assumption that consciousness can be ignored, which they, too, cannot prove. Why do we keep living with these unprovable X factors? Several answers come to mind.

1. **Faith:** We believe in certain things and that's good enough.

2. **Necessity:** We have to make sense of the world, even if there are glitches along the way.

3. Habit: The unprovable assumptions haven't bothered anybody so far, and therefore we've gotten into the habit of forgetting them.

4. Conformity: The system may be flawed, but everybody else uses it, so I will, too. I want to belong.

Lump all of these reasons together, and lesser mortals—even lesser mortals trained in science—find it easy to defend systems that have flaws they don't want to admit to. But it's not just the Achilles' heel in logic that plagues us. We are trapped by the implications of Gödel's second theorem, which holds that a logical system cannot reveal its inconsistencies; blindness is built in. I know that I am humanizing mathematics, which marks me as a total outsider, but systems engulf us at every turn—systems of politics, religion, morality, gender, economics, and above all, materialism. It's vital to know that you have been conditioned to accept these systems without regard for their unprovable assumptions. (Note that unprovable isn't the same as wrong. I can't prove that my mother loved me, but it's still true.)

Several times Leonard has asserted that we can't long for childish things like God, the afterlife, or the soul, and then expect them to be true. I don't think spirituality came about from wishful thinking. It came about because the world's sages, saints, and seers managed to escape the limitations of the logical system that Leonard has put so much faith in.

Gödel's insights can be extended to show us that logic machines can't make creative leaps, because any system that can't reveal its internal flaws will always be confined within the prison of its logic. Think of a computer that can detect a million shades of red. If you ask it which one is the nicest, it has nothing to say. "Nice" is outside its logic. Fortunately, Nature refuses to be imprisoned by logic, and we humans have taken our cue from that. When Picasso invented Cubism, when Tolstoy imagined Anna Karenina jumping in front of the train, when Keats wrote the final draft of "Ode to a Nightingale" in a frenzied few minutes, turning a promising poem into a masterpiece, creativity made leaps that were not based on mixing and matching the ingredients of what came before. Logic didn't come into it.

Leonard mentions Deep Blue, the chess-playing computer. On May 11, 1997, Deep Blue won a six-game match against the world chess champion, Garry Kasparov. This victory took ten years to achieve, growing out of a student project at Carnegie Mellon University. It was an anguishing emotional loss for Kasparov (we know the computer felt nothing about winning), who had defeated Deep Blue just the year before. But I'd like to turn this feat on its head. Deep Blue is a perfect example of a self-contained logical system that cannot escape its basic assumptions.

The machine knew nothing outside number crunching, and therefore it didn't know how to play chess at all. It only knew how to shuffle, at lightning speed, the human knowledge it was fed. Chess grandmasters display a lovely arrogance about what they do. Alexander Alekhine, a legendary Russian champion, was asked by awestruck admirers how many moves ahead he could look in a game. He replied coolly, "I can only see ahead one move, the right move." Chess playing is intuitive. It involves grasping the whole board, reading your opponent, taking risks, and so on. Grandmasters don't memorize thousands of games by rote to get where they are. They *learn* from thousands of games, which is entirely different. The mind is training the brain, which in turn gives the mind a higher platform to stand on, and thus the process continues, mind and brain evolving together. All that Deep Blue could do was to suck up this knowledge and spit it back out.

Finally, one branch of AI is devising artificial hands to replace hands lost in battle; countless disabled veterans and other amputees will benefit if the project succeeds. Figuring out the complex signals sent to and from a human hand is incredibly difficult. Could a prosthetic hand one day mold a beautiful sculpture like the Venus de Milo? Could it ever feel the cool hard surface of the marble? To oppose such altruistic work seems wrong, and critics of AI are routinely treated as enemies of progress. But we have to consider the work of a neuroscientist at the Salk Institute in San Diego, Vilayanur Ramachandran, and his amazing work with amputees.

After an amputation, many patients experience phantom limbs.

They continue to feel that the lost hand or arm is still there, and phantom limbs can be excruciatingly painful, often due to the sensation that the muscles are permanently clenched. Professor Ramachandran knew that drugs often do little for this pain, even strong doses of powerful painkillers. Pondering the problem, he made a creative leap. He took a patient whose right arm had been amputated and sat him in front of a box that had a mirror inside dividing the box in two. When the patient's left arm was placed in the box, he was asked to peer inside. What he saw were two arms, the right one being simply a reflection. But to the naked eye, the mirror image looked real.

The patient was then asked to clench and unclench both hands, the real and the phantom one. To the astonishment of everyone, this simple action could bring relief, sometimes instantaneously, to acute, intractable pain. The brain was fooled by the sight of a "real" right arm, and Ramachandran suggests that the area of the brain that received input from the limbs (the somatosensory cortex) had become cross-wired—it was mapping the lost arm by adapting other nearby regions reserved for the feet and face. Showing it the image of a right arm inside the mirror box enabled the brain to remap it, and thus unclenching the phantom muscles became possible. (A curious sidelight to Ramachandran's theory that the brain had become cross-wired is that sometimes the feelings from the amputated arm were transferred to the area that received sensations from the face. Thus, stroking a patient's face made him report that he felt the stroking on his lost arm.)

This could happen only because the mind, being different from the brain, figured out how to trick the brain and its pain signals. Ramachandran's methods are being tested in veterans' hospitals. Not all amputees benefit fully, and the amount of time spent in the mirror box varies, but the key thing was to prove that sudden change is possible. Neuroplasticity, the ability of old pathways to turn into new ones, took on new prestige.

I want to go a step further. If we could discover what's inside the mind, a door would open to higher intelligence. The trick—and it's the trick of all time—is that the mind can be explored only by the mind.

Every person knows how to look inside. We reflect, we second-guess, we try to make sense of our own motives. (A few familiar examples: "Why did I say something so stupid?" "I don't know how I knew, I just knew." "What made me eat the whole thing?") Knowing your mind isn't easy. The difference between a spiritual life and every other life comes down to this. In spirituality, you find out what the mind really is. Consciousness explores itself, and far from reaching a dead end, the mystery unravels. Then and only then does wisdom blossom. *The kingdom of God is within, I am the way and the life, Love thy neighbor as thyself*—these are not objective statements of fact. They cannot be deduced through computation. The mind has looked deeply into itself and discovered its source, which is transcendent.

Speaking of the presence of God, Hebrews 11:3 says, "What is seen was not made out of what is visible." If you want, you can match that statement with quantum physics, but in the end, it comes from something else, the ability of the mind to know itself. That, too, is an unprovable assumption, but what saves us is that this particular assumption is true.

15

Is the Universe Thinking Through Us?

DEEPAK

One of the most admired organizations in the world is Doctors Without Borders, whose courageous members travel to the world's trouble spots bringing healing. It would be inspiring if disputed borders gradually dissolved around the globe, but the most embattled borders are mental, and they are the ones that need to dissolve first. Everyone, even the most open-minded among us, is trapped behind such borders.

Let's say you are reading this chapter sitting outside under a tree on a sunny day. You lean back against its rough, cool bark to think. In order to have such thoughts you need red blood cells coursing through your bloodstream; that's how the brain gets the energy it needs to think. You also need the sun, without which life couldn't exist on Earth. You need the tree, because without photosynthesis, animals that breathe oxygen would never have appeared. Doesn't this mean that the tree and the sun are as much you as your blood? The boundaries that we've set up between mind, body, and the natural world are convenient, of course, and living within boundaries becomes second nature as we all learn to define ourselves as mothers, fathers, children, spouses, or singles when we get home. Yet the cosmos forgot to specialize, so it delivers reality all at once, in a huge messy package.

This fact can be overwhelming (which often drives people back to the comfort of their familiar pigeonholes). It implies that the universe—all of it, not just our cozy corner—is working through each of us. In order for you to take your next breath, the entire universe had to collaborate—you are the growing tip of the cosmos, the fresh spark of life being pushed forward by all that exists, the way the tip of a green needle on a Pacific sequoia is being pushed by the whole forest and ultimately the entire Earth.

Summon the courage to see yourself this way. Put aside any limited definition of who you are, and be without borders for a moment. I propose that it's not just the physical universe that is acting through you. When you pierce the mask of matter, you realize that the universe is also loving through you, creating through you, and evolving through you. Such a truth is very personal. To accept the spiritual life at all, this truth must be real for you, for it is the connection to a higher reality. Science sees human beings as isolated specks in the cosmos, accidental outcroppings of mind in a mindless creation. Yet mind is the connection that makes spirituality real. As the universe acts through you, it wraps you in the cosmic mind.

How do you know that you even have a mind? Without taking a philosophy course, most people intuitively accept René Descartes's maxim, "I think; therefore, I am." But they wouldn't say the same about a tree, a cloud, a neutron, or a galaxy. Borders are stubborn; walls are thick. What we need is a more unbounded definition of mind that embraces everything.

In his intriguing book *Mindsight,* Dr. Daniel Siegel, a deeply inquisitive psychiatrist from UCLA, provides just such a definition, and he has gone to the trouble of testing it out. At first he tried to define mind by asking various colleagues (all of them presumably had a mind), but no one could give him a satisfactory answer. Siegel was particularly interested in qualities of mind that could not be ascribed to the brain, and he found one: the mind's ability to observe. How we are able to observe the world is one of the greatest of all mysteries. If you try to claim that the brain is the same as the mind, you must answer a simple question: none of the ingredients in a brain cell—proteins, potassium, sodium, or water—can observe, but you can; so how did these objects acquire that capacity?

Let's have a writer explore the mystery in eloquent fashion: "I am a camera with its shutter open, quite passive, recording, not thinking. Recording the man shaving at the window opposite and the woman in the kimono washing her hair. Someday, all this will have to be developed, carefully printed, fixed." The setting is Nazi Germany. The speaker is the nameless narrator in Christopher Isherwood's haunting short novel *Goodbye to Berlin,* whose characters would become famous in the movie musical *Cabaret.* The speaker stands in for Isherwood himself, who wanted to keep truth alive by becoming an objective observer of history as Hitler plunged Europe into the horrors of a second world war. But certain facts work against Isherwood: the eye isn't a camera. The brain has no photographic images inside it. Perception is a function of consciousness, so the mind comes first, before any physical apparatus—eyes, ears, or brain. That's why Isherwood says "I" am a camera.

Clearly Leonard's basic allegiance lies with fixed mechanisms. He offers eye-catching optical illusions to prove that some things are seen the same way automatically, no matter how hard you try to see them a different way. To me, optical illusions prove exactly the opposite. Here's a classic example:

What do you see in this picture—a white vase standing in the middle of the image, or two faces in black silhouette looking at each other? Both are possible, and the whole point is that you can *decide*

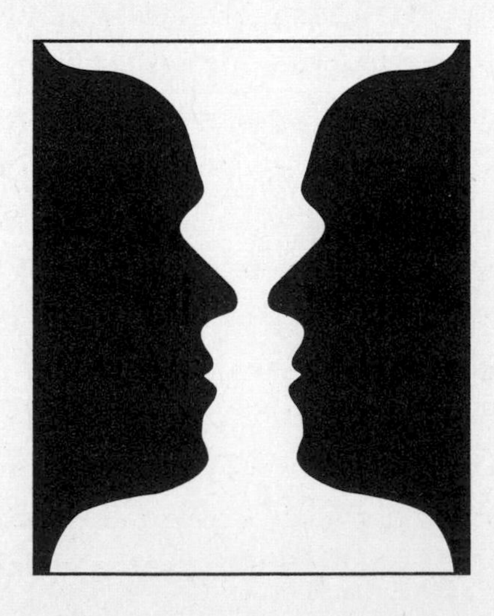

which one you want to see. You can switch from one to the other at will. As with every aspect of being an observer, the process is mental.

If perception came down to a physical mechanism, a camera, there would be no choice involved. The brain would take a snapshot, develop the image, and print it out. In fact, the brain does none of those things. It is set up solely to represent the mind, which sees, interprets, picks out details, chooses different viewpoints, etc. Presented with an optical illusion, your mind has the ability to see from at least two different points of view. For a second example, stare at the X in the diagram below.

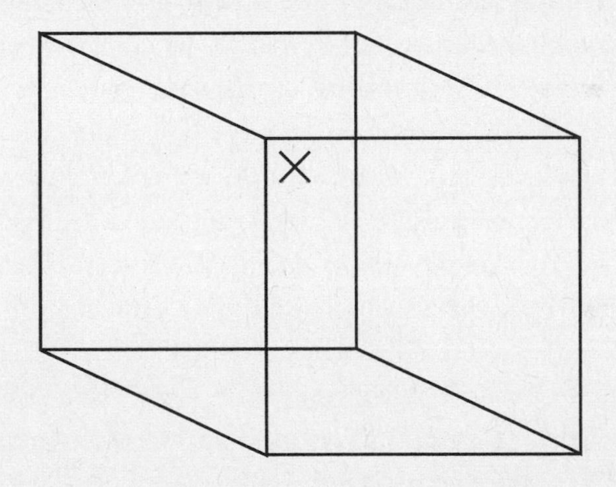

If you see the X as close to you, then it is on the front of the box. If you see the X as far from you, it is on the back of the box. You are the one making this choice; your brain isn't making it for you. Leonard's optical illusions were selected to force you to see in a fixed way. But that's because the brain is a fallible organ. For example, the visual cortex has a specific region devoted to recognizing faces, but it cannot recognize faces that are upside down. Try this yourself. Take a photo of a famous movie star and present it to a friend but turned upside down. Your friend won't be able to see that the photo is of Elizabeth Taylor or Robert Redford. However, the mind knows how to overcome this fallibility. It can look for clues, even in an upside-down photo—for example, spotting Bob Dylan's distinctive frowzy hair or

Captain Hook's eye patch. Then it becomes possible to override, at least partially, the limitations of a physical organ.

The brain can limit the mind, of course. If you should happen to have a blinding migraine or a brain tumor, you might not be able to see images at all. Certainly your visual cortex isn't set up to register ultraviolet or infrared light, as bees and snakes can. So physical limitations do count. They just don't provide proof of what the mind can and can't do.

Getting back to Daniel Siegel and his quest for a definition of mind, he made a good choice in settling on our ability to observe, and in particular on the mind's ability to observe itself. It's impossible to imagine a computer that can meditate and, by doing nothing else, arrive at insights and breakthroughs, much less change its wiring—but we can do both. Siegel eventually devised his own definition of mind, and having presented it to scientific audiences since 1993, he has met with no objection. The mind, he says, is an "embodied and relational process that regulates the flow of energy and information." That's a mouthful, but what makes Siegel's definition watertight is that not a single term can be omitted. Let's break it down one word at a time.

"Embodied": The mind makes itself known through an organ of the body, the brain.

"Relational": Our minds reflect the environment around us. We are constantly being shaped by the people around us, responding to their habits, speech, gestures, and facial expressions.

"Process": Mind is an activity. It isn't static but dynamic.

"Regulate": The jumble of data that the universe produces would be chaotic unless something organized it into a coherent reality. To keep reality intact, each part must be regulated with every other part.

"Flow": There is an uninterrupted stream of consciousness to parallel the uninterrupted stream of external events.

"Energy": To keep the flow going takes energy, at all levels from the immensity of the Big Bang to the micro level of ions passing through the cell membrane of a neuron.

"Information": Every piece of data can be seen as information, containing a bit of meaning.

What's so apt about these terms is that you can apply them to every aspect of Nature. As proud as we are of being human, mind is present in an amoeba, a mouse, a neuron, and a distant galaxy. Information and energy flow everywhere; they must be processed and distributed; their activity forms a tight web that connects everything in existence. As a universal definition of mind, this one is hard to improve upon.

Now we have a basis for asking if the universe is thinking through us, or to be more personal, through you. The answer is yes. It's such a simple answer that in my experience almost nobody resists it. In front of audiences I begin by pointing out that solid objects are deceptive. In reality, everything in the universe is a process with a beginning, a middle, and an end. "Photon" and "electron" aren't nouns, as far as Nature is concerned; they are verbs. Next I ask the audience to look at themselves.

Are you, too, a process in the universe, with a beginning, a middle, and an end? They nod yes.

Is your brain part of the process? Yes.

Is the electromagnetic storm in your brain giving rise to thoughts? Yes again, and we are almost there.

Then is the universe thinking through you? Most people have no trouble answering yes. If the universe can light up the sky with jagged arcs of lightning on a humid summer night, it can set up the lightning storms that appear on brain scans. In the present chapter all I've done is to define "thinking" as a process of mind rather than of brain, and most people don't object to this, either.

LEONARD

I grew up in an observant Jewish family, so I was surprised one day when my mother told me that she didn't believe in God. I asked her to explain, and she said that she used to believe, but she couldn't reconcile God with her experience of losing her family in the Holocaust. On bad days, I remember that I know what she means.

It was years ago, and I had just dropped my son Nicolai off for his fourth day of kindergarten. I stopped on my way to the subway to speak to another parent. I heard an odd sound and looked up the street to see a jumbo jet heading my way, but flying so low it felt eerie. A second or two later it passed overhead, seemed to bank slightly, and quietly penetrated the ninety-fifth floor of the north tower of the World Trade Center a short distance away. Almost immediately the upper floors spewed fire. The thunder of the crash came a half second later, as it would have if lightning had hit. The street broke into chaos, and the air filled with screams, and a rain of fiery debris. What haunts me most is the thought of the ninety-two people I saw obliterated in that moment—my involuntary feeling of connection with those people I never knew, but whose last moments staring out their windows in terror I can't stop imagining. Nicolai, his five-year-old face apparently pressed against the large picture window in his nearby classroom, saw it all, too, including those who leaped from the roof to avoid being incinerated.

Deepak wrote that we humans are "the growing tip of the cosmos, the fresh spark of life being pushed forward by all that exists," and that the universe is loving through us, creating through us. He says that "to accept the spiritual life at all," that truth must be real for us. In taking the point of view of science, and rejecting Deepak's version of spirituality, I sometimes find myself feeling like the haggard and unshaven Humphrey Bogart sending away beautiful Ingrid Bergman at the end of *Casablanca,* offering up my cold, calculated assessment that the problems of us little people—and our feelings—don't amount to a hill of beans in this crazy universe. But if Deepak is right about a universal consciousness, and that the universe is loving through us, then it must also be hating through us, murdering and destroying through us, doing all the things that humans do in addition to loving, including the acts that blew up my mother's faith in God. Deepak avoids talking about this dark side, but if the universe is working through each of us, then this universal connection must be a double-edged sword.

Though I believe in neither the God of the Bible nor the immaterial world Deepak advocates, I don't agree with him that to embrace the scientific view is to turn my back on spirituality. The great physicist Richard Feynman lost his childhood sweetheart, and the "great love" of his life, to tuberculosis when they were in their twenties, just a couple years after they had married. He once told me that he was not angry about it, because "you can't get angry at a bacteria." How rational and scientific, I remember thinking. But I later learned that he also wrote a letter to her—more than a year after her death:

> *D'Arline,*
>
> *I adore you, sweetheart. . . . It is such a terribly long time since I last wrote to you—almost two years but I know you'll excuse me because you understand how I am, stubborn and realistic; and I thought there was no sense to writing. But now I know my darling wife that it is right to do what I have delayed in doing, and what I have done so much in the past. I want to tell you I love you.*
>
> *I find it hard to understand in my mind what it means to love you after you are dead—but I still want to comfort and take care of you—and I want you to love me and care for me. . . .*

Richard Feynman was not only one of the greatest physicists in history, he was also famous among physicists for his passionate insistence that all theories be closely connected to experimental observations. Feynman felt lucky to have met his soul mate, even though he knew that what they felt for each other could be traced to physical processes, just as her death could be traced to a bacterium. And even though he knew she wasn't really there with him, he felt Arline's spirit through all the ensuing decades, until the day he himself died. That love is a mental phenomenon governed by the laws of nature he studied didn't diminish the depth of Feynman's feelings, or make him any less spiritual in his approach to life; and that he didn't really know what it meant to love Arline, or want her to love him, after her death, didn't cause him to deny that love. He knew that the effort to understand the mysteries of nature, of our mind and our existence, would never bring him in conflict with what he felt in his heart. Indeed, penetrating those mysteries is one of the ultimate triumphs of all the qualities that make us human.

As Deepak says, science draws borders; scientists believe it does that for good reason—to exclude from our worldview that which is not true. But there is plenty of room within those borders for emotion, and meaning, and spirituality. A scientific and a spiritual life can exist side by side.

Is the universe thinking through us? Even in our speculations, we scientists are careful. We want to see our ideas quoted in journals like *Physical Review* and *Nature,* not the *Encyclopedia of the Wrong.* As often happens when questions are posed in words rather than in precise mathematics, the scientific answer depends on the definition of terms. In chapter 14, I described the computational theory of mind. If by "thinking" one means, as some do, computing, then yes, the universe *is* thinking, because all objects follow mathematical laws and hence their behavior embodies the results of the computation dictated by those laws. Physicist Seth Lloyd wrote, "The universe is a quantum computer," and we are all part of it. In that sense I could agree with

Deepak that we are all part of a universal mind, and that the universe is thinking through us.

But when Deepak argues that the universe thinks through us, he means more than that. He sees us all as connected through a universal consciousness imbued with wonderful qualities such as love, and presumably also its opposite, hate. Embedded within this consciousness, somehow, are our immaterial minds, which express themselves through and also control our physical brains. As evidence of that view, he offers the faces/vase image below.

He says your ability to choose whether to see the two faces in black silhouette or the white vase is evidence that the mind is not a physical mechanism, for a physical mechanism can only "take a snapshot, develop the image, and print it out." He says that the nonphysical mind, in contrast, "interprets, picks out details, chooses different viewpoints, etc." But Deepak is wrong about your degree of control in the vase/faces illusion. You cannot choose to see either the vase or the faces. There is no immaterial mind that can overcome the structure of the physical brain.

Try it. If you keep looking long enough, you'll find that—whichever object you choose to focus on—your brain overrules your choice and flips a visual switch, so that now you see the other object. For example, if you focus on the vase, you cannot indefinitely make your mind consider the space around it as dead space, and not interpret it as two faces. Some people with mood disorders have very long switch periods, minutes compared to seconds, but everyone switches (researchers haven't relied on people's self-reports to know this; people's switching can be measured using external instruments).

Your visual experience when you look at a "bistable" image like this depends on many factors, such as conscious effort, prior exposure to the image, and its particulars, such as shading, but it also depends on limitations imposed by your physical brain. For example, scientists who studied people as they observed this image found that when subjects are experiencing the faces, but not the vase, a part of the temporal lobe that specializes in face recognition—the specialized region Deepak mentioned—is active. That area, called the fusiform face area, relies on the face being in a normal orientation, and as Deepak said, its efficacy is greatly diminished if you see a face that is, say, upside down. Flip a face over, and the hypothesized immaterial mind should not be fooled, but the physical brain *will* behave differently. So here is a test: look at the inverted faces/vase image below. Since your brain is in charge, you'll find the face images less conspicuous than before, but you will still switch perspectives.

In Deepak's other example he says that if you look at the X in the cube below, you are making the choice of whether to see it at the front or the back of the box. I disagree. So let's consider a simpler challenge. Armed with the conscious knowledge that the image below is not really a cube, but just some lines on a flat page, command your immaterial mind to take over from your physical brain. Try to perceive what you know are just meaningless lines on a page as nothing more than that. Can you look at the diagram below and *not* see a cube? If, as Deepak says, your brain is the mere servant of your mind, a camera or instrument your mind uses, while *you*—your mind—are really making the choice, you should be able to look at the diagram and not perceive a cube. But you can't.

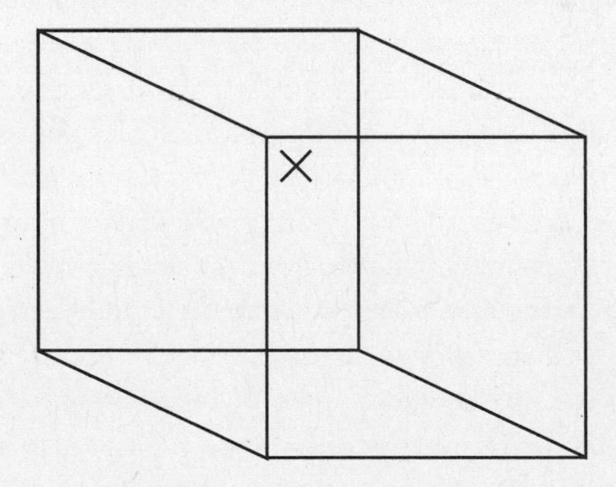

Deepak employs such examples when he thinks they support his argument and dismisses them when they don't, saying they arise because the mind is expressed through "a fallible organ." But that is just the point: scientists have been able to show that *every* aspect of human thought and behavior that has ever been studied is expressed through that fallible physical organ.

Everywhere we look we see evidence that the mind is a phenomenon of the brain. Daniel Siegel, a professor of psychiatry at UCLA, whose book *Mindsight* Deepak tries to use as evidence to the contrary, opens the book with a story that very clearly illustrates the physical basis of what we call "mind." It concerns a family in which the mother, a previously warm and loving presence, was in an automobile accident that severely damaged a part of the prefrontal cortex that is involved in "creating empathy, insight, moral awareness, and intuition." The result: a person who, while sane and rational and reasonably functional, felt no emotional connection to her family. As she herself described the difference she experienced in her new way of being: "I suppose I'd say that I've lost my soul."

Siegel was brought in to work with the family, for the children were very badly affected by the change in their mother's behavior toward them. He showed the family a scan of Barbara's brain and where it had been damaged, so that they would understand that "her brain was broken," as one of the children subsequently put it. But the other child, not satisfied with this explanation, said, "I thought love came from the heart." Siegel answered that she was right, and that the networks of cells around the heart and throughout the body communicate directly with the social part of our brain and "send that heartfelt sense right up to our middle prefrontal areas." But since that was the part of Barbara's brain that was damaged, it could no longer receive the signals. Over time, the family began to heal, but Barbara never did. Siegel wrote that the "damage to the front of her brain had been too severe, and she showed no signs of regaining her connected way of being." Her brain was broken, so her mind was, too.

The famous twentieth-century philosopher Bertrand Russell was once asked what he would say if he died and God confronted him, demanding to know why Russell had been an atheist. Russell famously said he would reply that it was God's fault. "Not enough evidence, God! Not enough evidence," he would say.

Deepak portrays the scientific insistence on data as cold and impersonal. I would be dishonest to dispute it when he says science sees human beings as "isolated specks in the cosmos, accidental outcroppings of mind in a mindless creation." There is much in humanity to be thankful for, but to deny that we are isolated specks in the cosmos is to avoid the truth rather than embracing it. Deepak said it takes courage to see ourselves as he suggests we should, but he paints a rosy picture, one that, as in the quote above, he likes to contrast with the worldview of science. What takes real bravery is to embrace the reality we actually observe, without regard to whether it is a bleak or a rosy picture. To grow old, to see friends die and planes crash, and to experience love and loss without the comforting illusion of a living, thinking universe imbued with a divine essence is what takes courage.

At the same time, I do choose a less bleak outlook. To me, though humans might be isolated specks with accidental outcroppings of mind, what is important is that we do have minds, we do have emotions, and we do have the capacity to experience art and beauty and joy. We are made according to chemistry and physics, but we are not "just" chemistry and physics. We are more than the sum of our components, and more than just alive. We are uncaring atoms and molecules that have banded together to care for one another, to feel love—and unfortunately also hate, as well as many other emotions, some exalted, some not. I feel connected. Tiny speck in the vast cosmos that I am, I feel a kinship with all the other tiny specks, and grateful for my brief moment of existence as a physical phenomenon, connected with all the other phenomena in nature. I am glad to be even a small part of an unthinking but wonderful and ever-changing universe.

PART FIVE

GOD

16

Is God an Illusion?

DEEPAK

There is no escaping the fact that the world isn't what it seems. Among the great quantum pioneers, I've mentioned Niels Bohr, who declared that what we accept as real is based on the unreal. In his Nobel Prize speech of 1932, Werner Heisenberg concluded that the atom "has no immediate and direct physical properties at all." The universe was doing a vanishing act at that time, and it certainly hasn't become more solid since. Mystery and wonder don't need the permission of science to exist, but in this case they got it.

That still leaves us far from God the benign creator, but spirituality isn't about defending the patriarchal God of convention. It's about a shift in consciousness. The famous Indian spiritual teacher J. Krishnamurti was speaking one afternoon against the backdrop of the Swiss Alps. Krishnamurti insisted on rigorous debate with his audiences—one of his main doctrines was that no one should blindly follow a guru or self-professed holy man—and on this occasion the exchange stalled out. People seemed baffled by the shift in consciousness that Krishnamurti wanted them to achieve.

Instead of going deeper, he turned toward a hazy peak in the distance. "If for one moment you could actually *see* that mountain," he murmured, "you would completely understand. Reality has been hidden from you, yet it is everywhere waiting to be noticed."

Unless you perceive things in a new way, spiritual teachings are a fiction. But how do those of us with normal perception arrive at such a state? How can we actually *see* the mountain? First, we need to realize what "normal" seeing actually means. In our everyday waking state we are:

1. Overshadowed by bodily sensations, information from our five senses, and past conditioning
2. Bound by the brain and its physical limitations
3. Dulled into following well-worn channels of perception, seeing the world today just as we saw it yesterday
4. Doubtful of our true purpose and destination
5. Haunted by fears and hidden memories from the past
6. Blind to what lies beyond the boundary that separates life and death

Seeing, it turns out, has problems. Happily, there are ways to escape these limitations of everyday consciousness. The world's religious scriptures are based on such journeys from the "normal" to the extraordinary, but so are inspired music, art, and poetry, not to mention the sudden spiritual experiences that life can bring to almost anyone. (Over half of people responding to a Gallup poll report that at least once in their lives they have seen light around another person, gone into the light themselves, felt the presence of the departed, or seen auras.) But science won't be satisfied until such experiences can be replicated, and whenever people "go into the light," a broad term for entering a higher state of consciousness, one thing is usually lacking: a way to repeat their journey afterward.

The spiritual life fills this void. It provides a path to higher consciousness that is universal. Let me try to offer a reliable map to such a journey, because this is one area where validation has taken place many thousands of times over the centuries.

Stage 1: The opening. We have a powerful personal experience that lifts us out of everyday consciousness. This can be a sudden insight that changes life forever or a taste of unity consciousness, or it can be something as simple as knowing that you are safe, or that everything in your life has a reason and a purpose.

Stage 2: Revising the meaning of life. Either slowly or all at once, we realize that material life is not what it seems on the surface. There is a higher purpose, which implies a mind or a consciousness, that is greater than the individual mind.

Stage 3: Becoming part of the plan. If higher reality begins to make more sense than everyday life, we begin to find ways to be transformed. Our desire to live on a different plane grows.

Stage 4: Walking the path. With a vision in mind, now we make the process of reaching higher reality a serious undertaking. The goal is God or higher consciousness, and a way must be found to get there.

Stage 5: Illumination. Higher consciousness becomes a living reality. The shift is complete. We know no other way to see the world than as an aspect of the divine. Indeed, sacred and nonsacred no longer have separate meanings. There is only the light of consciousness everywhere we look.

I believe you can take any deeply meaningful life and find that it fits this template, without regard to religion. In fact, one of the greatest failings of religion is the claim that it has patented the way to God. In the West we sorely lack a nonreligious model for becoming illuminated, but we're getting there. Ironically, we can thank science for forcing us to drop preconceived notions and rely on hard evidence. Reason tells us that the Buddha, Saint Paul, Bernadette of Lourdes, and Sri Ramakrishna underwent a common experience, and like scientists sorting out how an apple and a rose are linked to the same genus, we can fit unique examples of spiritual awakening into the same template.

Thoreau writes about "the solitary hired man on a farm in the outskirts of Concord, who has had his second birth" and who muses that his "peculiar religious experience" might not be true. This leads to a passage in *Walden* that has haunted me for decades, ever since I first read it:

> Zoroaster, thousands of years ago, travelled the same road and had the same experience, but he, being wise, knew it to be universal, and treated his neighbors accordingly, and is even said to have invented and established worship among men.

We may smile at the naïveté of dating religion to one seer in ancient Persia and his "peculiar religious experience." Zoroaster, who was born somewhere between the eighth and tenth centuries BCE, was a relative late-comer compared to the Vedic seers of India. But I share Thoreau's essential point, as well as what he advises the hired man on the farm outside Concord to do:

> Let him humbly commune with Zoroaster then, and, through the liberalizing influence of all the worthies, with Jesus Christ himself, and let "our church" go by the board.

In contemporary language this means that the person who has had a sudden awakening should see himself or herself in the great tradition of awakening. The second reference, to putting the church aside, has already occurred on a large scale.

Your life and your mind sit somewhere on the continuum of awakening, even if we have turned our backs on religion en masse. The conscious process of coming to terms with higher reality is personal, spontaneous, and never on schedule. Countless people have revised their view of material life and decided to walk the spiritual path—but then they stop. Sadly, as long as divinity means the God of organized religion, the spiritual path has little chance of going mainstream. Faiths promote their own agendas. They want followers who pose no doubts. They insist that their dogmas were handed down by God, even when history reveals that they were devised by powerful clerics. So many agendas work against finding the divine that the situation has given rise to a cynical joke: God handed down the truth, and the Devil said, "Let me organize it."

Yet the spiritual path exists, and can be followed. Once you stop seeking the traditional God, a different goal takes its place: transcendence. To transcend means to go beyond. The process should be regarded as natural; in fact, we transcend all the time. When a three-year-old throws a tantrum until it gets its way, the mother doesn't sink to the level of childish demands. She knows that something else is at work: her child is tired or upset or anxious. She transcends the context created by the tantrum and goes to a different level of experience. The Buddha and Jesus did much the same thing. In the context of a suffering, bewildered humanity, they didn't recommend pleasure as a replacement for pain. They pointed to solutions that went beyond the level of the problem. Without transcendence, our experience of suffering will never change.

Pioneering American psychologist William James reduced the mystery of finding God to a simple statement: "All around us lie infinite worlds, separated only by the thinnest veils." The secret is that these veils are made of consciousness that is blocked and constricted, while the other worlds are made of consciousness that is expanded and free. The spiritual path is about removing the veils over your own awareness, which requires dedication. What makes the effort worthwhile is knowing that awareness can grow at any moment.

A simple parable comes to mind. In a remote town lived a gifted sculptor. His work decorated the town's streets and parks, and everyone agreed that it was extraordinarily beautiful. But the artist was reclusive and remained out of sight. One day a visitor arrived and so admired the statues that he insisted on meeting the sculptor, but no one could tell him how to find the artist he sought. In fact, it turned out that none of the townspeople had actually met him; the sculpture had just appeared, as if on its own.

Then an old man stepped forward and said that he had been fortunate enough to meet the elusive sculptor.

"How did you manage that?" the visitor asked.

The old man replied, "I stood before these wonderful works of art and kept admiring them. The more I gazed, the more I saw. There was intricacy and subtlety beyond anything I had ever observed before. I couldn't stop marveling. Somehow the sculptor must have become aware of my rapture, for, to my astonishment, he appeared by my side.

"I said, 'Why did you pick me to show yourself to, when no one else has found you, no matter how hard they searched?'

"He said, 'No creator can resist appearing when his work is loved as intensely as you love mine.'"

In this little tale we can see the only article of faith that is necessary. If you go deeply enough into your own awareness, you will find a place of silence and peace. In time this place will reveal much more than that, however. The source of creation abides there, and the more you experience it, the richer and more beautiful creation becomes. Beyond suffering lies bliss; transcending takes you to the domain of light. Go there and find out for yourself, not by seeking God but by seeking reality.

Eventually the artist won't be able to resist you—your appreciation of what he has created will draw him to your side. Then the divine will no longer be a projection or a fantasy. It won't be a wished-for father or a mother. The props of wish-fulfillment will no longer be needed once you come face-to-face with your own inner experience of the divine. You won't care about such things as worldviews. They are only stepping-stones for the mind. In the end it is irrelevant whether the nameless assumes the face of God or not. Far better is reality itself, seen as clearly as the light of day.

LEONARD

Decades ago, when I was still walking to school every day carrying a lunch bag, I decided that physical science held the key to the mysteries I wanted to understand—both those of the universe around me (what caused the sun to shine, the stars to twinkle above me, the elegant butterfly to look the way it did), and those of my own mind. Over thousands of hours I digested lectures, articles, and books, and over thousands more I explored the cosmos through mathematics. Could I ever understand everything? Or anything? What does it even mean to understand?

In college my friends and I believed in a hierarchy of truths, like the layers of the Earth's atmosphere. Mathematics formed the outer and most sacred sphere of the truths—the heavens, the realm of pure ideas. Just below it was the stratosphere, consisting of theoretical physics, the fundamental truths of everything palpable. The less rarefied regions beneath were where we located the applied sciences, thick, turbulent, and polluted with endless facts and intricate details. Philosophy, metaphysics, and theology were hard to place, however. Our attitude toward those subjects varied with the philosopher, with the particular work we were reading, with our mood, and even with how much we had been drinking. Baruch Spinoza, the great rationalist philosopher and lens grinder, for example, wrote a book called *Ethics,* an unforgiving seventeenth-century critique of traditional religion and morality. It seemed heavenly in its mathematical structure of definitions, axioms, propositions, and proofs, but disappointing in the less-than-mathematically-precise arguments Spinoza attached to that formal structure. My friends and I found we could feast on his ideas, but afterward be uncertain what we had eaten. So in the end we had sympathy for Spinoza's approach, but skepticism about whether he had made a convincing case for his ideas.

The disciplines of science and mathematics were different. We reveled in their precision. We celebrated the methodologies they had developed to avoid the pitfalls of human bias and subjectivity. And, knowing both how they reached their conclusions and how open they were to changing those conclusions when the evidence warranted, we felt confident that we could trust what they told us.

Though many today debate the validity of "mere scientific theories," the same people depend upon science, in all aspects of their lives, without giving it a second thought. The power of the scientific method is why advertisers scream that their detergents have been "scientifically" proven to remove stains, while no one would spend a dime to advertise that metaphysics proves their mouthwash will sweeten your breath. Doubters appear on television and radio to deny the reality of evolution or the Big Bang theory, but somehow, when the debate gets reduced to coffee stains on white dress shirts, or how to treat pneumonia, they find reality versus illusion easy to sort out, and they side with the scientists.

No one, of course, employs the scientific method to sort out truth from illusion regarding their own lives. You might think that the person you married is your ideal partner, but that person would frown on it if you married a dozen others to gather evidence for your theory. You might think your great talents guaranteed your professional success, but you won't restart your career to test that hypothesis. You might believe in an afterlife, but you're in no rush to perform the one experiment that could tell you if you are right. We build our worldviews through experience, intuition, schooling, books, and dialogues with people whose ideas we trust and respect. We make decisions about what is true and what is false, but most of us rarely think about how we come to our beliefs. We assume we're rational—and therefore correct—and then we rush off to our next appointment. But there are factors affecting our beliefs that most of us are unaware of. These manifest most visibly when it comes to issues of great personal importance. It is well known amongst psychologists, for example, that the burden of proof people ordinarily require varies with the desirability of what is being "proved"—and that it is our subconscious mind that adjusts the dial.

There are many examples of this subconscious activity in the scientific literature. Studies show that it takes an immense body of smoking-gun evidence to convince us we are nitwits, but to persuade us of our genius or talent requires only the flimsiest data; that partisans studying research on a political issue will view the same methodology as lacking or sound, depending on whether the conclusion implied conforms to their beliefs; that jurors tend to discount shoddy evidence of guilt when they have sympathy for a defendant, but judge it as convincing when they consider the defendant unlikable. In one study researchers presented two groups of volunteers with documents adapted from a murder trial in which both the prosecution and the defense presented strong arguments. The documents included parts of the trial transcript, and a newspaper article from the time of the trial that was neutral with regard to whether the defendant had committed the crime. But the two groups were shown slightly different articles. The newspaper article shown to one group quoted neighbors describing the defendant as an unpleasant character. When asked whether they thought—based on the trial transcript—that the prosecution had proved the defendant's guilt, the subjects who were led to believe the defendant was unlikable were much more apt to conclude that the prosecution had indeed proved its case.

In all these instances, people thought they were being objective, but their objectivity was an illusion. In truth, our everyday analyses always depend on prior beliefs and desires. If we want to reach a certain conclusion, our brains will alter the way we perceive and weigh data and analyze arguments. And—most important—our brains do this beneath our level of awareness. So it is often possible to honestly believe what we wish to believe, even though an objective observer would come to another conclusion. Psychologists sometimes call this motivated reasoning, which is a force to be reckoned with when examining why we might choose to believe in a seductive worldview involving universal consciousness and a loving universe.

Consider Deepak's interpretation of the experience of "going into the light" or of seeing an aura around someone. According to a study in the British medical journal *The Lancet,* about 10 percent of cardiac patients resuscitated from clinical death report either "out-of-body" or "near-death" experiences. How should we interpret them? Deepak associates such experiences with entering a "higher state of consciousness." That explanation fits neatly into Deepak's worldview, which, like Buddhism, postulates an immaterial mental realm. But is that just a desirable way to interpret the event, or is there evidence to support that view? Through painstaking effort, and the application of new technologies for examining the brain, scientists have been studying such events and coming to a very different conclusion. For example, David Comings, a neuroscientist who specializes in altered states of consciousness, has found that near-death experiences seem to come when the brain is deprived of oxygen for prolonged periods of time, immediately prior to brain damage. Out-of-body experiences, too, seem to have a physical basis. That was dramatically illustrated recently by the case of a forty-three-year-old woman who reported feeling a "lightness," and said she was floating about two meters above the bed, close to the ceiling, and seeing herself, from above, lying in bed. She wasn't near death, but rather had had electrodes implanted in a part of her temporal lobe called the right angular gyrus. The electrodes were part of a treatment for severe epileptic seizures, but they also allowed researchers to probe the effects of mild electrical stimulation on the brain. As professional skeptic Michael Shermer reported in *The Believing Brain,* researchers found that, by varying that stimulation, they could not only induce out-of-body experiences, but actually control the height above the bed at which the patient reported floating.

Richard Dawkins wrote that, when watching a great magic trick, it's hard not to think, "It must be a miracle," even though in that case one knows full well it isn't. It is even more difficult not to believe in the miraculous when we have a vested interest in an interpretation that science contradicts. Exotic and poorly understood phenomena like out-of-body experiences can be a haven of "proof" for ideas that have been examined in better-understood contexts, and found lacking. But even if a phenomenon is as yet poorly understood, it is useful to recall that throughout history, the inexplicable has repeatedly been found, over the long run, to have a natural explanation. Never, so far, has a scientist been forced to fill in a gap in understanding with Sidney Harris's famous cartoon caption "Then a miracle occurred."

We might have good objective reasons for the views we hold so dearly, and we might not, but either way it is best to be able to assess how convincing the evidence is. This is not always easy. If you ask a friend why she believes in God, or in a higher presence, she probably won't say she came to that belief through a series of controlled experiments. More likely, she'll say she feels it, or she just knows. Is God merely an illusion perceived by those who are looking for a divine presence? Science is the best method we know for discovering truth about the material universe, but the powers of science are not without limit. Science does not address the meaning of life, nor can it, for now, explain consciousness. And science will never be able to explain *why* the universe follows laws. So while science often casts doubt on spiritual beliefs and doctrine insofar as they make representations about the physical world, science does not—and cannot—conclude that God is an illusion.

Since Deepak offers parables, I, too, will offer an illustrative story, this one symbolic but true. In 1969 Richard Feynman invented a model of hadrons—particles like the proton and the neutron that interact via a force called the strong force, which, as its name indicates, is the most potent force in nature. In Feynman's model, a hadron is like a bag containing partons that move freely inside it, but are constrained not to leave the bag. Feynman used his parton picture to explain certain data regarding what happens when hadrons are smashed into one another at high energy, and it worked well, which is to say its predictions were confirmed. Yet, since partons must stay within the bag—within the hadron—we don't see partons. Are they real, or just an illusion, mere constructs in Feynman's intellectual model? It's a metaphysical question, and though Feynman once famously said he "was under doctor's orders not to discuss metaphysics," he did address it. He wrote that since partons help us make sense of what is going on, they may be a useful "psychological guide," and "if they continue to serve this way to produce other valid expectations they would of course begin to become 'real,' possibly as real as any other theoretical structure invented to describe nature."

Useful as these "psychological guides" are in physics, there is no reason not to employ similar guides in our spiritual life, so long as they help us to make sense of the universe—and are compatible with our observations of it. Many people intuitively believe in a higher power, and draw solace, strength, and courage from that belief. When faith feels real to a person, and when that particular belief does not lead to conflict with what we observe in the physical world, there is nothing science says to oppose it. If, however, we are asked to believe in a God who created the universe a few thousand years ago, and we have convincing evidence that the universe is much older than that, then we have a conflict. But the demands of science do not preclude the rewards of spirituality. In fact, even Albert Einstein, almost superhuman in his clarity of thought and his ability to reason, exulted in his sense of spiritual connection to the universe. In his case, it was the very "rationality" of that universe that shaped his spiritual life:

> *Whoever has undergone the intense experience of successful advances made in [science], is moved by profound reverence for the rationality made manifest in existence. By way of the understanding he achieves a far-reaching emancipation from the shackles of personal hopes and desires. . . . And so it seems to me that science . . . contributes to a religious spiritualization of our understanding of life.*

17

What Is the Future of Belief?

DEEPAK

I equate the future of belief with the future of God. Modern belief in a deity is much diluted, which requires some blunt talk. All too easily discussions of God descend into polite murmurs over tea and cookies about matters that have no bearing on the practicalities of everyday life. For countless people, personal belief is both embarrassing and shaky. I've been advocating for the spiritual path, on the other hand, as something vital and urgent. The future of the planet depends upon raising our consciousness. Since God is intimately tied into who we are and what life means, there is no separate future for God and for the individual. You and I will make decisions that determine if God has a viable tomorrow.

The main issue is a shift away from God as an external force to God as an inner experience, from religion to spirituality. We are not talking about a return to mysticism. Modern life rests upon two things: information and personal satisfaction. There are no concrete facts, however, to support that Jesus Christ rose from the dead, that the angel Gabriel dictated the Koran, or that Moses actually existed. That basically leaves personal satisfaction, and here spirituality finds its entry point.

People crave meaning and value in their lives. If an inner experience of God can fulfill this craving, it will supplant the old ways of approaching the divine. An external God sitting above the clouds, as represented in popular religion, faces bleak prospects. Behind every pulpit an invisible clock is ticking, counting off the hours as thousands of people flee from churches and temples. In almost every developed country religious attendance has waned to no more than 20 percent of the population, and in many places, such as Scandinavia, the figure is less than 10 percent. God is no longer personally satisfying. Religions emphasizing sin, guilt, and punishment are not likely to attract people who want to pursue fulfillment without being stigmatized (one example being the Catholic Church's condemnation of Eastern meditation as a way to experience the divine, which is viewed as heretical).

I am convinced that a shift inward is necessary. We must free ourselves from the burden of religious dogma, but at the same time we can't give in to materialism. Even when espoused by a voice as sympathetic as Leonard's, mechanistic determinism offers no personal satisfaction, except for a certain grim appreciation of the courage required to face a universe that is cold and void. Spirituality can do better. However, skeptics have a right to ask for specifics, and there are certainly pitfalls that must be avoided.

A visitor once came calling on a famous spiritual teacher. He was motioned to sit on the floor in a cool, empty room. Across from him the teacher, dressed in white, sat silently while an attendant poured tea. It was difficult for the visitor to wait; he was obviously agitated.

Once the attendant had left, the visitor burst out, saying, "Sir, I hear that you are revered and wise. But I have met many others just like you, and frankly, it has taken me a long time to convince myself that I should even tell you my problem. You are likely to fail me, as everyone else has."

The teacher looked unperturbed. "What is this problem of yours?"

The visitor sighed. "I am sixty years old, and ever since childhood I've been drawn toward God. While earning a living and raising a family, I also undertook an intense quest. I've prayed, meditated, and gone on retreats. I've read every scripture. I have passed months in the company of so-called holy men."

"And what did your quest reveal? Did you not find God?"

The man shook his head mournfully. "I've had countless experiences that seemed right. I've had visions. I've been filled with the light.

Every golden bell and Buddha you can imagine has appeared to me. But it has all turned to dust. I feel empty and depressed, abandoned by God. It's as if I've experienced nothing."

"Of course," the spiritual teacher murmured.

The man looked startled. "You mean there is no God?"

"I mean that the mind can project whatever you ask it to. If you are looking for golden Buddhas, they will appear. So will all the gods, or *the* God. Each path leads to a goal that is known in advance. But is that really God? God is about freedom. You have ardently pursued all these disciplines, yet you have not arrived at your destination." Then the teacher smiled enigmatically. "Now let me ask you one question: can you discipline yourself to be free?"

This exchange, which happens to be true, casts radical doubt upon conventional paths to God. But it also points to another way, sometimes called "the pathless path." On the pathless path there is no fixed goal and no prescribed process to follow. Looking at yourself intimately, from moment to moment, you peel away the unreal aspects of yourself, until only the real is left. Many things are unreal, as viewed by the wisdom traditions of the world. Ignorance is unreal, especially ignorance about who you really are. The ego and its urgent needs are unreal. Since these needs form the foundation of most people's lives, you can see that deep transformation is called for.

Getting there sounds forbidding, I know. Having said farewell to organized religion, is it any better to be faced with your own pain and suffering? Can anyone really give up the ego's endless desires? The saving grace of the spiritual path is that it comes naturally. Although life is full of suffering and the ego demands to be satisfied, those things are not as substantial as they seem. If you walk through a garden rank with weeds and withered flowers, they look real enough. But looks are deceiving—the deeper reality is the garden's rich soil and the renewal of life, which cannot be stopped. In our case, the nurturing soil is the soul, and the renewal of life happens within. You don't have to tell your body to renew itself; it does so naturally. You don't have to force your mind to have new perceptions; billions of bits of sensory data flood the mind every day. The process of renewal guides life on every level. To me, a viable future for the spirit centers on discovering that the creative and evolutionary impetus in Nature is the same force that resides at the heart of who we are.

I've often thought that everyone would lead a spiritual life if they simply watched young children closely. Children don't resist their inner development. It doesn't frighten them that life may stop at three years old or five or ten; when it's time to give up paper dolls and learn to read, this new stage emerges spontaneously. How do three-year-olds prepare to be four? They don't. Each child does what he or she does, just allowing whatever comes next to unfold naturally. This is a secret that Nature has mastered—how to allow the new to emerge, not by destroying the old, but by welling up from the inside, invisibly and silently, until the new has flowered of its own accord.

On the pathless path a similar process takes place. New qualities arise in your awareness, not by warring against your old self but by encouraging natural growth from within. Modern people may be baffled as they look back at the age of faith, but the fact that we live in a different age doesn't mean that spiritual awakening is invalid. Quite the opposite, in fact. Cleared of the undergrowth of dogma and superstition, the spiritual path has become much easier to walk. The best way to fulfill your aspirations is by waking up, and instead of choosing to renounce the world in God's name, choosing to embrace it in your own. However, to make such a radical shift possible, we must explore what it means to wake up.

The process of waking up centers on transcending, as we've discussed. Beyond our everyday waking state we find a deeper level of inner silence. This is not a search for peace and quiet; rather, we are transcending the maelstrom of everyday thoughts to find the source of the mind. Practically speaking, there are many levels of transcendence. The most profound is deep meditation, which is known to alter brain structure and lead to lasting transformation. At the shallow end there is the exhilaration that fans feel at a football game, or that serious shoppers find when they snap up a bargain. These two poles seem to have nothing in common, but there is a hidden bond. Whenever you experience any quality of pure consciousness, however fleeting, you have transcended.

Pure consciousness isn't a way of thinking or a point of view. It's the unseen potential from which everything springs. The qualities of pure consciousness seem subtle at first, but they grow more powerful as you proceed farther on your path. Here are the chief qualities described in the great wisdom traditions.

Ten Qualities of Pure Consciousness

1. Pure consciousness is ***silent*** and ***peaceful.*** When you experience this quality, you are free of inner conflict, anger, and fear.

2. Pure consciousness is ***self-sufficient,*** or centered within itself. When you experience this quality, the need for distraction vanishes. You are comfortable simply being here. The mind is not restless in its quest for stimulation.

3. Pure consciousness is fully ***awake.*** This quality is experienced as mental alertness and freshness. The mind is no longer dull or fatigued.

4. Pure consciousness contains ***infinite potential;*** it is open to any outcome. When you experience this quality, you are no longer bound by fixed habits and beliefs. The horizon seems open, the future full of possibilities. The greater your experience of pure potential, the more creative you become.

5. Pure consciousness is ***self-organizing.*** It effortlessly coordinates all aspects of existence. You are experiencing this quality when things fall into place of their own accord. There is less struggle to force different parts of life to harmonize, because you are more in tune with the natural harmony that runs through everything.

6. Pure consciousness is ***spontaneous.*** Timetables, boundaries, and rules don't apply; nor are they needed. Breaking free of old constraints, whatever they may be, makes you feel safer about expressing who you are and what you want without constraints. This is the state of absolute freedom, which you experience whenever you feel liberated.

7. Pure consciousness is ***dynamic.*** Although not in motion, it provides energy for all the activity in the universe. You experience this quality when you feel that you can fully embrace life. You have the energy and the will to do great things.

8. Pure consciousness is ***blissful.*** This is the root of happiness and its highest expression. Any surge of happiness, whatever the cause, is a taste of bliss. An orgasm is blissful, but so is compassion. Every experience of love can also be traced back to its origins in bliss.

9. Pure consciousness is ***knowing.*** It contains the answers to all questions and, more crucially, the practical knowledge needed for unfolding the universe, the human body, and the mind. Any experience of intuition, insight, or truth taps into this quality.

10. Pure consciousness is ***whole.*** It is all-encompassing. Therefore, despite the infinite diversity of the physical world, at a deeper level only one process is occurring: wholeness is moving like a single ocean that holds every wave. You experience this quality when your life makes sense and you feel a part of Nature; you are at home simply by being alive.

As you can see, I haven't used any religious terms, yet this is divinity, stripped of the demands of faith and obedience. At this point you cannot be expected simply to accept that these ten qualities are, in essence, divine. However, you can use this idea as a working hypothesis.

In that sense you are the experimenter and the experiment. If you are transcending everyday reality, these ten qualities will grow in your life. You will notice more fulfillment and creativity. Your sense of being secure will grow as you come to know who you really are.

Now we can say with certainty what kind of action the spiritual path calls for. You don't have to brace yourself to become "spiritual" in quotation marks. The only requirement is that you measure your activity, inner and outer, by one criterion: does it develop the qualities of pure consciousness? In spirituality there is room for deeply religious people and room for worldly people (including scientists). Doing good works and being of service aren't a guarantee that you are transcending, yet they are landmarks on a recognized spiritual path, and countless seekers do find that service increases their sense of bliss, peace, centeredness, and self-sufficiency. Another recognized path is deep contemplation; yet another is mindfulness—becoming aware that your thoughts are just thoughts, coming and going like clouds against the eternal sky of consciousness. The spiritual experiment is yours to set up as you wish.

I'm not suggesting that you take on a regimen and cling tightly to it. Consciousness does the work for you here, just as genes do the work for an embryo as it develops. The difference is that spiritual growth requires choice. As you come to know what pure consciousness is, you orient your mind toward it. To avoid sounding too mystical, let me share with you a parable from the Upanishads of ancient India.

A coachman is driving a team of horses, using his whip to goad them faster and faster. It's a sunny day; he feels exhilarated, as if he owns the world. From the inside of the coach a faint voice says, "Stop." In his excitement the coachman ignores the voice; he's not even sure he heard anything. Again the same softly spoken command comes from inside the coach: "Stop."

This time the coachman knows that he has heard the command, which makes him angry, so he flogs the horses to race even faster. But the voice from inside the coach continues to repeat its command, never raising its voice, until the coachman remembers something. His passenger is the owner of the coach! The coachman pulls on the reins, and slowly, slowly the horses come to a halt.

In the parable the horses are the five senses and the mind, constantly being whipped onward by the ego. The ego feels that it controls everything. But the owner of the coach is the soul, whose soft voice waits patiently to be heard. When it is, the ego relents. It gives up false ownership. The mind slows down its frantic activity, and in time it learns to stop. Stopping isn't an end unto itself; it's the basis for knowing who you really are: a soul with all its divine attributes. Those attributes are the qualities of pure consciousness.

I believe every home should have a nook devoted to divinity—a shrine of roses, or an altar of scented lavender. A shard of crystal would do, or a small bronze Buddha placed where the sun can warm it. We need daily reminders if the divine is to have a future. Reminders of what? The voice from inside the coach.

I won't cramp the soul by attempting to define it. That's part of the experiment, to find out for yourself. But I can't resist sharing a passage from the *Bhagavad Gita,* written from the soul's point of view:

> *This entire universe is pervaded by Me, the unmanifest Brahman. All beings depend on Me. I am the origin, the seed of all beings.*
>
> *There is nothing, animate or inanimate, that is not pervaded by Me. I am found in all of creation. I am inside and outside all that exists.*

In the end, the spiritual path does one simple thing: it makes those timeless words come true for you. Belief becomes knowledge that can be trusted, and on that basis God can once again be revered.

LEONARD

Auguste Comte, one of the most influential French philosophers of the first half of the nineteenth century, wrote extensively on the nature of knowledge, what it means, and how we obtain it. Alas, Comte chose an unfortunate example to illustrate his philosophy, based on what he considered an infallible scientific fact: "On the subject of stars . . . we shall not at all be able to determine their chemical composition or even their density. . . . I regard any notion concerning the true mean temperature of the various stars as forever denied to us." Just fourteen years later Gustav Kirchhoff and Robert Bunsen discovered that we could indeed determine the properties of stars by analyzing the light they emit, and today we use that method, spectroscopy, to measure chemical abundances, temperatures, density, and many other properties of distant planets, stars, and galaxies. Some of the astronomical objects we study in this manner are over ten billion light-years away.

According to the dictionary, the difference between knowledge and belief is that belief implies confidence, while knowing implies certainty. Though there are issues of consistency, and philosophers may debate the issue, it is possible to achieve certainty of sorts in mathematics—you apply the rules and derive the consequences, an exercise in pure logic. But in our everyday lives, and even in science, that distinction between what we "know" and what we merely "believe" is difficult, or even impossible, to make. We might think that we can distinguish between *believing* we won't get sick from the raw halibut at our local sushi bar, and *knowing* that tomorrow the sun will rise in the east. But can we really? We base what we think we know—the beliefs we feel certain about, or which we at least don't question—partly on empirical evidence. We have seen or heard about the sun rising on every other day of our lives, and even before our birth, so we "know" it will rise again tomorrow. In 1812 astronomer and mathematician Pierre-Simon Laplace employed probability theory to examine the degree of certainty that is justified in that prediction, based solely on the fact that the sun had risen every day for the past five thousand years (the approximate age of the Earth according to biblical accounts). He came up with odds of 1,826,214:1 in favor. But empirical evidence is not all we use to form our beliefs, and Laplace pointed out that people probably have a much higher confidence that the sun will rise than his calculation indicates because they know that the laws of nature—technically, gravity and celestial mechanics—call for it to do so. Ironically, today's theories of physics tell us the sun probably won't forever continue to rise or even to exist. As I said earlier, in roughly seven billion years the sun will grow 250 times larger (and 2,700 times more luminous) than today, ballooning out to fill the entire sky, and then probably swallowing up the Earth. Billions of years later it will burn out and shrink and turn into a kind of stellar corpse called a white dwarf. In a sense all that we say we "know"—except perhaps mathematical truths—is just belief, and so the question of the future of theological belief is tied to how and why we believe things in ordinary life, and even in science.

Bertrand Russell wrote that "believing seems the most mental thing we do." It is also one of the most complex and varied things we do. Not only do observation, theoretical understanding, our needs, desires, and biases, our emotions and mood, and our existing framework of beliefs all interact in a complex manner to affect the way we form beliefs; but we might not even be aware of what our beliefs are, because while we may consciously think we believe one thing, on a deeper unconscious level we may believe and sometimes act on the opposite belief. For example, consider an experiment concerning what psychologists call the illusion of control, the unconscious belief that we are the masters of our own fate even when we consciously know we aren't. In the study, employees at an insurance agency and a manufacturing company on Long Island who donated $1 for an office lottery were either allowed to choose their own lottery ticket, or else given one at random by the seller. Then the morning of the drawing the sellers approached each buyer individually and said that someone else "wanted to get into the lottery, but since I'm not selling tickets anymore, he asked me if I'd find out how much you'd sell your ticket for. It makes no difference to me, but how much should I tell him?" Though it is doubtful that many of the subjects *consciously* believed they had any skill in picking the winner of a random drawing, they seemed to believe it nonetheless: those who had been randomly assigned a lottery ticket agreed to sell it back for an average price of $1.96, while those who had chosen their own ticket demanded on average $8.67. Our inner weighing of evidence is not a careful mathematical calculation resulting in a probabilistic estimate of truth, but more like a whirlpool blending of the objective and the personal. The result is a set of beliefs—both conscious and unconscious—that guide us in interpreting all the events of our lives.

For example, a parent's suggestion to a teenager that he or she put on a jacket before going out in the cold can be construed as an attempt to exercise control, as a protective move arising from an exaggerated fear of sickness, or as an expression of love and concern. A computer analyzing only the parent's words might make no inference, or might request more data. But the teenager on the receiving end will probably jump to some conclusion based upon his or her prior beliefs regarding the parent, and not give much thought to alternative possible interpretations. Like Comte, we assume we know.

Our brains, for good reason, tend to jump to conclusions based on past experience, rules of thumb, and an existing framework of beliefs. We wouldn't get very far in life if before setting out to watch the beauty of the sun rising, we debated whether it was likely to do so. In fact, evolution favored those whose gut reactions guided their choices. When the earth starts trembling and you're standing at the foot of a cliff, it's better to run first and engage in making theories about what is happening later. If instinct hadn't made a connection between cause and effect and catalyzed an immediate plan of action in response, our ancestors would have been devoured while still pondering that mysterious movement in the bushes. As William James remarked, "The intellect is built up of practical interests."

Whatever the future of theological belief, people will always adhere to belief systems that gratify their emotional needs. None of us can function without having faith of one sort or another. Entrepreneurs start businesses on faith, immigrants with no concrete prospects move to a new country on faith, writers toil for long hours in the faith that people will want to read their words. There are atheists who put their faith in lucky numbers, and otherwise rational lawyers who eat tuna, a cheeseburger, or a Mayan sun salad each day that a trial continues because they think it's their lucky meal. "You certainly wouldn't want to learn that your heart surgeon or your 747 pilot always wears the same pair of underwear when it's time to perform," said an attorney critical of such practices, but there are no doubt surgeons and pilots who do just that. There was even a politician in Israel who was famous for always wearing his lucky underwear on election day. Physicist George Gamow told a story about Niels Bohr, who supposedly had a horseshoe nailed over the door of his country cottage. When asked how a famed scientist can have faith in a charm, Bohr replied that he didn't believe in it, but "they say that it does bring luck even if you don't believe."

We call these superstitions, but they reflect a deep emotional need to justify believing that when we undertake a great challenge, we will succeed. William James wrote about imagining himself stuck in the Alps, in a position from which the only escape is by a terrible leap. "Being without similar experience," he wrote, "I have no evidence of my ability to perform it successfully, but hope and confidence in myself make me sure I shall not miss my aim, and nerve my feet to execute what without those subjective emotions would perhaps have been impossible. But suppose, on the contrary . . . I feel that it would be sinful to act upon an assumption unverified by previous experience; why, then I shall hesitate so long that at last, exhausted and trembling, launching myself in a moment of despair, I miss my foothold and roll into the abyss." James wrote that "every philosopher, or man of science either, whose initiative counts for anything in the evolution of thought, has taken his stand on a sort of dumb conviction that the truth must lie in one direction rather than another . . . and has borne his best fruit in trying to *make* it work."

Without their faith, many theoretical physicists facing years sequestered in dank offices, working on complex calculations with no promise of success, might indeed have insufficient courage to leap across their abyss. For example, one of the central pursuits in fundamental physics today is the quest for an ultimate and elegant theory that unifies all four of the forces we have observed to operate in nature. One of the forces, gravity, is known to obey this simple equation, which was formulated by Einstein:

$$R_{\mu\nu} - \frac{1}{2}g_{\mu\nu}R = \frac{8\pi G}{c^4}T_{\mu\nu} - g_{\mu\nu}\Lambda$$

$$G = 6.67300 \times 10^{-11}\ \mathrm{m^3/kg{-}s^2}\,;\ c = 299{,}792{,}458\ \mathrm{m/s}$$

Einstein's equation is of course not really as simple as it looks—it takes much study to be able to learn to apply it, and to understand what it means, and it is one of the most difficult equations in all of physics to solve. But it has a simple physical interpretation, and is a highly economical way of expressing a complex thought through mathematics, with the left side of the equation representing the structure of space-time, while the right side represents its matter-energy content. To a physicist, that makes it an elegant equation. Now have a look at the current theory of the other three forces, called the "standard model." It doesn't matter what the symbols here actually mean, for even an uninformed viewer will be able to see that this set of symbols is quite a bit messier and less elegant than the one above:

$$\mathcal{L} = -\frac{1}{4}B_{\mu\nu}B^{\mu\nu} - \frac{1}{8}tr(\mathbf{W}_{\mu\nu}\mathbf{W}^{\mu\nu}) - \frac{1}{2}tr(\mathbf{G}_{\mu\nu}\mathbf{G}^{\mu\nu}) \qquad \text{(U(1), SU(2) and SU(3) gauge terms)}$$

$$+(\bar\nu_L, \bar e_L)\,\tilde\sigma^\mu iD_\mu \begin{pmatrix}\nu_L\\ e_L\end{pmatrix} + \bar e_R\sigma^\mu iD_\mu e_R + \bar\nu_R\sigma^\mu iD_\mu\nu_R \qquad \text{(lepton dynamical term)}$$

$$-\frac{\sqrt2}{v}\left[(\bar\nu_L, \bar e_L)\,\phi M^e e_R + \bar e_R M^{e*}\bar\phi\begin{pmatrix}\nu_L\\ e_L\end{pmatrix}\right] \qquad \text{(electron, muon, tauon mass term)}$$

$$-\frac{\sqrt2}{v}\left[(-\bar e_L, \bar\nu_L)\,\phi^* M^\nu \nu_R + \bar\nu_R M^{\nu*}\phi^T\begin{pmatrix}-e_L\\ \nu_L\end{pmatrix}\right] \qquad \text{(neutrino mass term)}$$

$$+(\bar u_L, \bar d_L)\,\tilde\sigma^\mu iD_\mu \begin{pmatrix}u_L\\ d_L\end{pmatrix} + \bar u_R\sigma^\mu iD_\mu u_R + \bar d_R\sigma^\mu iD_\mu d_R \qquad \text{(quark dynamical term)}$$

$$-\frac{\sqrt2}{v}\left[(\bar u_L, \bar d_L)\,\phi M^d d_R + \bar d_R M^{d*}\bar\phi\begin{pmatrix}u_L\\ d_L\end{pmatrix}\right] \qquad \text{(down, strange, bottom mass term)}$$

$$-\frac{\sqrt2}{v}\left[(-\bar d_L, \bar u_L)\,\phi^* M^u u_R + \bar u_R M^{u*}\phi^T\begin{pmatrix}-d_L\\ u_L\end{pmatrix}\right] \qquad \text{(up, charmed, top mass term)}$$

$$+\overline{(D_\mu\phi)}D^\mu\phi - m_h^2[\bar\phi\phi - v^2/2]^2/v^2, \qquad \text{(Higgs dynamical and mass term)}$$

$$+\text{(Hermitian conjugate of some terms).} \qquad (1)$$

where $\bar\psi = \psi^\dagger$, and the derivative operators are

$$D_\mu\begin{pmatrix}\nu_L\\ e_L\end{pmatrix} = \left[\partial_\mu - \frac{ig_1}{2}B_\mu + \frac{ig_2}{2}\mathbf{W}_\mu\right]\begin{pmatrix}\nu_L\\ e_L\end{pmatrix}, \quad D_\mu\begin{pmatrix}u_L\\ d_L\end{pmatrix} = \left[\partial_\mu + \frac{ig_1}{6}B_\mu + \frac{ig_2}{2}\mathbf{W}_\mu + ig\mathbf{G}_\mu\right]\begin{pmatrix}u_L\\ d_L\end{pmatrix}, \qquad (2)$$

$$D_\mu\nu_R = \partial_\mu\nu_R, \quad D_\mu e_R = [\partial_\mu - ig_1B_\mu]\,e_R, \quad D_\mu u_R = \left[\partial_\mu + \frac{i2g_1}{3}B_\mu + ig\mathbf{G}_\mu\right]u_R, \quad D_\mu d_R = \left[\partial_\mu - \frac{ig_1}{3}B_\mu + ig\mathbf{G}_\mu\right]d_R, \qquad (3)$$

$$D_\mu\phi = \left[\partial_\mu + \frac{ig_1}{2}B_\mu + \frac{ig_2}{2}\mathbf{W}_\mu\right]\phi. \qquad (4)$$

ϕ is a 2-component complex Higgs field. Since $\mathcal{L}$ is $SU(2)$ gauge invariant, a gauge can be chosen so ϕ has the form

$$\phi^T_\bullet = (0, v+h)/\sqrt2, \qquad <\phi>^T_0 = \text{(expectation value of } \phi) = (0, v)/\sqrt2, \qquad (5)$$

where v is a real constant such that $\mathcal{L}_\phi = \overline{(\partial_\mu\phi)}\partial^\mu\phi - m_h^2[\bar\phi\phi - v^2/2]^2/v^2$ is minimized, and h is a residual Higgs field. B_μ, $\mathbf{W}_\mu$ and $\mathbf{G}_\mu$ are the gauge boson vector potentials, and $\mathbf{W}_\mu$ and $\mathbf{G}_\mu$ are composed of 2×2 and 3×3 traceless Hermitian matrices. Their associated field tensors are

$$B_{\mu\nu} = \partial_\mu B_\nu - \partial_\nu B_\mu, \quad \mathbf{W}_{\mu\nu} = \partial_\mu\mathbf{W}_\nu - \partial_\nu\mathbf{W}_\mu + ig_2(\mathbf{W}_\mu\mathbf{W}_\nu - \mathbf{W}_\nu\mathbf{W}_\mu)/2, \quad \mathbf{G}_{\mu\nu} = \partial_\mu\mathbf{G}_\nu - \partial_\nu\mathbf{G}_\mu + ig(\mathbf{G}_\mu\mathbf{G}_\nu - \mathbf{G}_\nu\mathbf{G}_\mu). \qquad (6)$$

The non-matrix $A_\mu, Z_\mu, W^\pm_\mu$ bosons are mixtures of $\mathbf{W}_\mu$ and B_μ components, according to the weak mixing angle θ_w,

$$A_\mu = W_{11\mu}sin\theta_w + B_\mu cos\theta_w, \qquad Z_\mu = W_{11\mu}cos\theta_w - B_\mu sin\theta_w, \qquad W^+_\mu = W^{-*}_\mu = W_{12\mu}/\sqrt2, \qquad (7)$$

$$B_\mu = A_\mu cos\theta_w - Z_\mu sin\theta_w, \quad W_{11\mu} = -W_{22\mu} = A_\mu sin\theta_w + Z_\mu cos\theta_w, \quad W_{12\mu} = W^*_{21\mu} = \sqrt2\,W^+_\mu, \quad sin^2\theta_w = .2315(4). \qquad (8)$$

The fermions include the leptons e_R, e_L, ν_R, ν_L and quarks u_R, u_L, d_R, d_L. They all have implicit 3-component generation indices, $e_i=(e,\mu,\tau)$, $\nu_i=(\nu_e,\nu_\mu,\nu_\tau)$, $u_i=(u,c,t)$, $d_i=(d,s,b)$, which contract into the fermion mass matrices $M^e_{ij}, M^\nu_{ij}, M^u_{ij}, M^d_{ij}$, and implicit 2-component indices which contract into the Pauli matrices,

$$\sigma^\mu = \left[\begin{pmatrix}1&0\\0&1\end{pmatrix}, \begin{pmatrix}0&1\\1&0\end{pmatrix}, \begin{pmatrix}0&-i\\i&0\end{pmatrix}, \begin{pmatrix}1&0\\0&-1\end{pmatrix}\right], \quad \tilde\sigma^\mu = [\sigma^0, -\sigma^1, -\sigma^2, -\sigma^3], \quad tr(\sigma^i) = 0, \quad \sigma^{\mu\dagger} = \sigma^\mu, \quad tr(\sigma^\mu\sigma^\nu) = 2\delta^{\mu\nu}. \qquad (9)$$

The quarks also have implicit 3-component color indices which contract into $\mathbf{G}_\mu$. So $\mathcal{L}$ really has implicit sums over 3-component generation indices, 2-component Pauli indices, 3-component color indices in the quark terms, and 2-component $SU(2)$ indices in $(\bar\nu_L, \bar e_L), (\bar u_L, \bar d_L), (-\bar e_L, \bar\nu_L), (-\bar d_L, \bar u_L), \bar\phi, \mathbf{W}_\mu, \binom{\nu_L}{e_L}, \binom{u_L}{d_L}, \binom{-e_L}{\nu_L}, \binom{-d_L}{u_L}, \phi$.

The electroweak and strong coupling constants, Higgs vacuum expectation value (VEV), and Higgs mass are,

$$g_1 = e/cos\theta_w, \quad g_2 = e/sin\theta_w, \quad g = 3.892e, \quad v = \sqrt{2}\cdot 180\,GeV = 254\,GeV, \quad m_h \sim 115 - 180GeV? \tag{10}$$

where $e=\sqrt{4\pi\alpha\hbar c}=\sqrt{4\pi/137}$ in natural units. Using (4,5) and rewriting some things gives the mass of A_μ, Z_μ, $W^\pm_\mu$,

$$-\frac{1}{4}B_{\mu\nu}B^{\mu\nu}-\frac{1}{8}tr(\mathbf{W}_{\mu\nu}\mathbf{W}^{\mu\nu})=-\frac{1}{4}A_{\mu\nu}A^{\mu\nu}-\frac{1}{4}Z_{\mu\nu}Z^{\mu\nu}-\frac{1}{2}\mathcal{W}^-_{\mu\nu}\mathcal{W}^{+\mu\nu}+\begin{pmatrix}\text{higher}\\ \text{order terms}\end{pmatrix}, \tag{11}$$

$$A_{\mu\nu}=\partial_\mu A_\nu-\partial_\nu A_\mu, \quad Z_{\mu\nu}=\partial_\mu Z_\nu-\partial_\nu Z_\mu, \quad \mathcal{W}^\pm_{\mu\nu}=D_\mu W^\pm_\nu-D_\nu W^\pm_\mu, \quad D_\mu W^\pm_\nu=[\partial_\mu \pm ieA_\mu]W^\pm_\nu, \tag{12}$$

$$D_\mu <\phi>_0=\frac{iv}{\sqrt{2}}\begin{pmatrix} g_2W_{12\mu}/2 \\ g_1B_\mu/2+g_2W_{22\mu}/2\end{pmatrix}=\frac{ig_2v}{2}\begin{pmatrix} W_{12\mu}/\sqrt{2} \\ (B_\mu sin\theta_w/cos\theta_w+W_{22\mu})/\sqrt{2}\end{pmatrix}=\frac{ig_2v}{2}\begin{pmatrix} W^+_\mu \\ -Z_\mu/\sqrt{2}\,cos\theta_w\end{pmatrix}, \tag{13}$$

$$\Rightarrow \quad m_A=0, \quad m_{W^\pm}=g_2v/2=80.425(38)GeV, \quad m_Z=g_2v/2cos\theta_w=91.1876(21)GeV. \tag{14}$$

Ordinary 4-component Dirac fermions are composed of the left and right handed 2-component fields,

$$e=\begin{pmatrix}e_{L1}\\ e_{R1}\end{pmatrix},\ \nu_e=\begin{pmatrix}\nu_{L1}\\ \nu_{R1}\end{pmatrix},\ u=\begin{pmatrix}u_{L1}\\ u_{R1}\end{pmatrix},\ d=\begin{pmatrix}d_{L1}\\ d_{R1}\end{pmatrix}, \quad \text{(electron, electron neutrino, up and down quark)} \tag{15}$$

$$\mu=\begin{pmatrix}e_{L2}\\ e_{R2}\end{pmatrix},\ \nu_\mu=\begin{pmatrix}\nu_{L2}\\ \nu_{R2}\end{pmatrix},\ c=\begin{pmatrix}u_{L2}\\ u_{R2}\end{pmatrix},\ s=\begin{pmatrix}d_{L2}\\ d_{R2}\end{pmatrix}, \quad \text{(muon, muon neutrino, charmed and strange quark)} \tag{16}$$

$$\tau=\begin{pmatrix}e_{L3}\\ e_{R3}\end{pmatrix},\ \nu_\tau=\begin{pmatrix}\nu_{L3}\\ \nu_{R3}\end{pmatrix},\ t=\begin{pmatrix}u_{L3}\\ u_{R3}\end{pmatrix},\ b=\begin{pmatrix}d_{L3}\\ d_{R3}\end{pmatrix}, \quad \text{(tauon, tauon neutrino, top and bottom quark)} \tag{17}$$

$$\gamma^\mu=\begin{pmatrix}0 & \sigma^\mu\\ \tilde{\sigma}^\mu & 0\end{pmatrix} \quad \text{where } \gamma^\mu\gamma^\nu+\gamma^\nu\gamma^\mu=2Ig^{\mu\nu}. \quad \text{(Dirac gamma matrices in chiral representation)} \tag{18}$$

The corresponding antiparticles are related to the particles according to $\psi^c=-i\gamma^2\psi^*$ or $\psi^c_L=-i\sigma^2\psi^*_R$, $\psi^c_R=i\sigma^2\psi^*_L$. The fermion charges are the coefficients of A_μ when (8,10) are substituted into either the left or right handed derivative operators (2-4). The fermion masses are the singular values of the 3×3 fermion mass matrices M^ν, M^e, M^u, M^d,

$$M^e=\mathbf{U}^{e\dagger}_L\begin{pmatrix}m_e&0&0\\0&m_\mu&0\\0&0&m_\tau\end{pmatrix}\mathbf{U}^e_R,\ M^\nu=\mathbf{U}^{\nu\dagger}_L\begin{pmatrix}m_{\nu_e}&0&0\\0&m_{\nu_\mu}&0\\0&0&m_{\nu_\tau}\end{pmatrix}\mathbf{U}^\nu_R,\ M^u=\mathbf{U}^{u\dagger}_L\begin{pmatrix}m_u&0&0\\0&m_c&0\\0&0&m_t\end{pmatrix}\mathbf{U}^u_R,\ M^d=\mathbf{U}^{d\dagger}_L\begin{pmatrix}m_d&0&0\\0&m_s&0\\0&0&m_b\end{pmatrix}\mathbf{U}^d_R, \tag{19}$$

$$m_e=.510998910(13)MeV,\quad m_{\nu_e}\sim.001-2eV,\quad m_u=1.5-3.3MeV,\quad m_d=3.5-6MeV, \tag{20}$$

$$m_\mu=105.658367(4)MeV,\quad m_{\nu_\mu}\sim.001-2eV,\quad m_c=1.16-1.34GeV,\quad m_s=70-130MeV, \tag{21}$$

$$m_\tau=1776.84(17)MeV,\quad m_{\nu_\tau}\sim.001-2eV,\quad m_t=169-174GeV,\quad m_b=4.13-4.37GeV, \tag{22}$$

where the **U**s are 3×3 unitary matrices ($\mathbf{U}^{-1}=\mathbf{U}^\dagger$). Consequently the "true fermions" with definite masses are actually linear combinations of those in $\mathcal{L}$, or conversely the fermions in $\mathcal{L}$ are linear combinations of the true fermions,

$$e'_L=\mathbf{U}^e_Le_L,\ e'_R=\mathbf{U}^e_Re_R,\ \nu'_L=\mathbf{U}^\nu_L\nu_L,\ \nu'_R=\mathbf{U}^\nu_R\nu_R,\ u'_L=\mathbf{U}^u_Lu_L,\ u'_R=\mathbf{U}^u_Ru_R,\ d'_L=\mathbf{U}^d_Ld_L,\ d'_R=\mathbf{U}^d_Rd_R, \tag{23}$$

$$e_L=\mathbf{U}^{e\dagger}_Le'_L,\ e_R=\mathbf{U}^{e\dagger}_Re'_R,\ \nu_L=\mathbf{U}^{\nu\dagger}_L\nu'_L,\ \nu_R=\mathbf{U}^{\nu\dagger}_R\nu'_R,\ u_L=\mathbf{U}^{u\dagger}_Lu'_L,\ u_R=\mathbf{U}^{u\dagger}_Ru'_R,\ d_L=\mathbf{U}^{d\dagger}_Ld'_L,\ d_R=\mathbf{U}^{d\dagger}_Rd'_R. \tag{24}$$

When $\mathcal{L}$ is written in terms of the true fermions, the **U**s fall out except in $\bar{u}'_L\mathbf{U}^u_L\tilde{\sigma}^\mu W^\pm_\mu\mathbf{U}^{d\dagger}_Ld'_L$ and $\bar{\nu}'_L\mathbf{U}^\nu_L\tilde{\sigma}^\mu W^\pm_\mu\mathbf{U}^{e\dagger}_Le'_L$. Because of this, and some absorption of constants into the fermion fields, the parameters in the **U**s are entirely contained in only four components of the Cabibbo-Kobayashi-Maskawa matrix $\mathbf{V}^q=\mathbf{U}^u_L\mathbf{U}^{d\dagger}_L$ and four components of $\mathbf{V}^l=\mathbf{U}^\nu_L\mathbf{U}^{e\dagger}_L$. The unitary matrices $\mathbf{V}^q$ and $\mathbf{V}^l$ are often parameterized as

$$\mathbf{V}=\begin{pmatrix}1&0&0\\0&c_{23}&s_{23}\\0&-s_{23}&c_{23}\end{pmatrix}\begin{pmatrix}e^{-i\delta/2}&0&0\\0&1&0\\0&0&e^{i\delta/2}\end{pmatrix}\begin{pmatrix}c_{13}&0&s_{13}\\0&1&0\\-s_{13}&0&c_{13}\end{pmatrix}\begin{pmatrix}e^{i\delta/2}&0&0\\0&1&0\\0&0&e^{-i\delta/2}\end{pmatrix}\begin{pmatrix}c_{12}&s_{12}&0\\-s_{12}&c_{12}&\\0&0&1\end{pmatrix},\quad c_j=\sqrt{1-s_j^2}, \tag{25}$$

$$\delta^q=57(14)\,\text{deg},\quad s^q_{12}=0.2243(16),\quad s^q_{23}=0.0413(15),\quad s^q_{13}=0.0037(5), \tag{26}$$

$$\delta^l=?,\quad s^l_{12}=0.57(3),\quad s^l_{23}=0.7(1),\quad s^l_{13}=0.0(2). \tag{27}$$

$\mathcal{L}$ is invariant under a $U(1)\otimes SU(2)$ gauge transformation with $U^{-1}=U^\dagger$, $detU=1$, θ real,

$$\mathbf{W}_\mu\rightarrow U\mathbf{W}_\mu U^\dagger-(2i/g_2)U\partial_\mu U^\dagger,\ \mathbf{W}_{\mu\nu}\rightarrow U\mathbf{W}_{\mu\nu}U^\dagger,\ B_\mu\rightarrow B_\mu+(2/g_1)\partial_\mu\theta,\ B_{\mu\nu}\rightarrow B_{\mu\nu},\ \phi\rightarrow e^{-i\theta}U\phi, \tag{28}$$

$$\begin{pmatrix}\nu_L\\e_L\end{pmatrix}\rightarrow e^{i\theta}U\begin{pmatrix}\nu_L\\e_L\end{pmatrix},\ \begin{pmatrix}u_L\\d_L\end{pmatrix}\rightarrow e^{-i\theta/3}U\begin{pmatrix}u_L\\d_L\end{pmatrix},\quad \begin{matrix}\nu_R\rightarrow\nu_R, & u_R\rightarrow e^{-4i\theta/3}u_R,\\ e_R\rightarrow e^{2i\theta}e_R, & d_R\rightarrow e^{2i\theta/3}d_R,\end{matrix} \tag{29}$$

and under an $SU(3)$ gauge transformation with $V^{-1}=V^\dagger$, $detV=1$,

$$\mathbf{G}_\mu\rightarrow V\mathbf{G}_\mu V^\dagger-(i/g)V\partial_\mu V^\dagger,\ \mathbf{G}_{\mu\nu}\rightarrow V\mathbf{G}_{\mu\nu}V^\dagger,\ u_L\rightarrow Vu_L,\ d_L\rightarrow Vd_L,\ u_R\rightarrow Vu_R,\ d_R\rightarrow Vd_R. \tag{30}$$

To both the trained and the untrained eye, the standard model is ugly—more like a circuit diagram for some high-tech appliance than an expression of simple physical principles. Yet it works very well. Is there a more elegant theory of these forces yet to be discovered? Richard Feynman said, "People say to me, 'Are you looking for the ultimate laws of physics?' No, I'm not. I'm just looking to find out more about the world. If it turns out there is a simple, ultimate law which explains everything, so be it; that would be very nice to discover. If it turns out it's like an onion, with millions of layers, and we're sick and tired of looking at layers, then that's the way it is." But despite Feynman's skepticism, if you asked those working in the field, you'd have trouble finding one who doesn't have faith that a more attractive theory does exist. Physicists take comfort in their faith that at its core, nature is simple and elegant. For them as for everyone else, belief based on feeling, desire, need, or intuition is a fundamental feature of the human mind.

Whenever we face difficulties, challenges, or uncertainties, it can be helpful to hold beliefs that stretch beyond that which we know without question to be true. Faith, as James put it, can be a great "working hypothesis." This is as true for scientists as it is for anyone else. In fact, it's important for scientists to formulate such working hypotheses (and then to be willing to jettison them if they aren't borne out) because if we didn't, we would never move forward in our knowledge of the universe. But working hypotheses like Deepak's, which insist on the primacy of an immaterial world, or like the beliefs of those who deny evolution or embrace supernatural miracles, are at odds with our knowledge of the world, often in active conflict with the physical laws that govern it. Hence they are flawed.

I agree with Deepak that it would be nice if over time theological belief shifted away from God as an external force that created and rules the universe to God as an inner experience. But God the Ruler has a long history. The strong human desire to understand the universe, and to attribute causes to the events that transpire in our world, gave rise, in ancient times, to myths, beliefs synthetically constructed to explain situations people simply didn't understand. The attraction of those myths was not so much in any objective truth they codified, but rather in their ability to provide comforting answers to the question "How did we get here and why?" Before the advent of science, God the Ruler was the answer. God the Ruler met other human longings, too—satisfying our need to believe that events happen for a purpose; that the world is just; that death is not the end, but a beginning.

Many predict the demise of this kingly and personal God as future science produces triumph after triumph. But science has already shown its prowess in the physical world—from demonstrating that the Earth is round to explaining how space is curved. We have seen evolution studied down to the molecular level, the universe explored almost back to the Big Bang, bacterial life synthesized, lambs cloned, surgery performed with lasers, people sent to the moon, robots sent to Mars, three-dimensional images of our brains, quantum teleportation . . . and still, the enthusiasm for religious explanations of the physical world remains strong.

The science of the future might produce a laser that teleports a synthetic lamb to Mars to feed robot astronauts, but there is no reason to think that that or any other spectacular feat would bolster the prestige of science at the expense of religious belief. If there's one area in which we might find ourselves in agreement with Mahmoud Ahmadinejad, the president of Iran, it's expressed in a letter he wrote to George W. Bush in 2006, saying that "whether we like it or not, the world is gravitating towards faith in the Almighty."

Sure enough, a Gallup poll taken not long after Ahmadinejad wrote his letter showed that 94 percent of Americans believe in God, 82 percent say religion is at least fairly important to them, and 76 percent say the Bible is the actual or inspired word of God. If those numbers are down, they certainly haven't fallen far. To believe is human, and belief in the traditional God seems to be alive and well, enjoying the likelihood of a long-lasting future.

18

Is There a Fundamental Reality?

LEONARD

On December 17, 1999, a woman known as F.B. in the neuroscience literature suffered a stroke on the right side of her brain. As a result, she lost the feeling of sensation on the left side of her body and could no longer move her left arm or leg, nor see anything on the left side of her visual field. And though F.B.'s memory was not tested, in patients with similar damage to the brain, access to memories of the left side of the world were found to be obliterated, presumably because the retrieval of those memories entails activation of some of the same neural circuits that are active during the actual perception of a scene.

When F.B. was asked to touch her left hand, she was unable to find it, and when it was pointed out to her, she denied it was hers. F.B. was fully oriented, did well on a mental-state examination doctors administered, and did not show any sign of general mental deterioration. But, as regards her left hand, she was seriously misinformed: she insisted it belonged to her niece.

The phenomenon of a patient disowning a limb was first documented in 1942. The delusion was named "somatophrenia." What is striking about somatophrenia is that patients suffering from it are profoundly unaware of their delusion, and maintain their belief regardless of strong evidence to the contrary. When pressed, they typically concede that what they are reporting sounds odd, but offer evidence to support their story. How could an otherwise intelligent and grounded individual hold such an absurd belief? Presented with a limb they can neither move, feel, nor remember, these patients' brains attempt to construct a coherent story that will lead to an apparently reasonable conclusion: that the limb does not belong to them. From the point of view of people who have normal brains, the patients' conclusions are flawed because of damage to both their sensory data systems and the particular brain structures that interpret that data. But even healthy human brains have limitations and peculiarities of design, and so healthy people, too, are constrained in the way they observe and interpret the world.

It would be narrow-minded for us to believe that our picture of the world is the definitive one. Aliens with senses and brains that function differently from ours might consider our perceptions to be as deluded as we believe F.B.'s to be. Or if they have superior brains, they might wonder about our primitive worldview the way we might wonder about that of a grasshopper or a bat. Yet we are as sure of the validity of our interpretation of reality as somatophrenics are of theirs.

Most people are what philosophers call "naive realists." They believe that an objective external reality exists, and is populated by objects with definite properties they can identify and codify. Experiments in psychology support the idea that people automatically assume their subjective experience to be a faithful representation of the real world. But long before knowledge of syndromes like somatophrenia, or access to technologies like fMRI that allow us to peer into the brain, there have been thinkers who have mounted impressive arguments against the beliefs of naive realism. For example, in 1781, German philosopher Immanuel Kant postulated that the reality we experience is one that has been constructed and shaped by our minds, minds limited by our beliefs, feelings, experiences, and desires.

In the century after Kant, developments in physics grew to require, more and more, the consideration of reality on a new level, beyond that which we experience in everyday life. Unseen entities such as electric and magnetic fields, atoms and electrons, began to creep into the intellectual theories of the physicists. Einstein would come to call the idea of the field "probably the most profound transformation which has been experienced by the foundations of physics since Newton's time"; Feynman would have a similar opinion about the concept of the atom. These were mental models. Physicists found them useful in analyzing the phenomena they studied, helping them visualize the events they observed, and hence aiding their ability to reason about them, and to make new predictions. But they were outside our usual experience, and, initially at least, not seen in the laboratory, either, so it was not clear to what extent they should be considered real. As Ludwig Boltzmann, the nineteenth-century physicist who might be called the father of modern atomic theory, wrote, such theories could be regarded as "merely a mental picture of phenomena that is related to them in the same way in which a symbol is related to the thing symbolized." In other words, atoms and fields were a kind of language.

Galileo said, "The universe is a grand book written in the language of mathematics," and that grand book has been the concern of scientists ever since. But are we reading the grand book of the universe, or are we writing one?

Mathematician David Ruelle, in a series of articles beginning with "Conversations on Mathematics with a Visitor from Outer Space," pointed out that humans do math (and therefore physics) with parts of their brain that evolved for other purposes. Our mathematical thinking, he says, is limited by poor memory and attention span, and by our peculiarly human reliance on visualization. This suggests that at least as far as the scientists constructing new theories of the universe are concerned, the innate peculiarities of the human style of theorizing must be added to the list of the influences affecting our concept of reality.

Consider, for example, the idea of the atom. In our everyday world we experience gaseous matter through its bulk characteristics such as pressure, temperature, and flow. Scientists had previously noted relations among those properties, but pioneers like Boltzmann realized that they could derive those conclusions from a model in which gases are made of atoms. The atomic model explains the properties of gases in terms of these hypothetical unseen entities. More important, the atomic picture could also be used to predict new phenomena. Many scientists opposed such theories on the ground that atoms were mere mathematical constructs and didn't "really exist." Then in 1905 Einstein showed that atomic and molecular processes accounted for the quantitative features of a phenomenon called Brownian motion, which is visible under a microscope. That was enough for most physicists to consider atoms to be real. But it wasn't until 1981 that scientists first "saw" a molecule "directly." And even then, what the scientists really did was to compile an image by scanning a needle over the surface of a material. So although some would say this constitutes "seeing" a molecule "directly," others would say it is merely a human artist-scientist's visualization of Boltzmann's mathematical construct, the "atom."

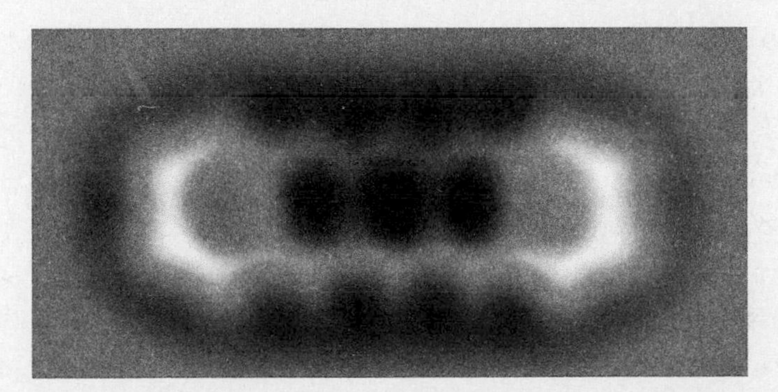

This image of pentacene, a molecule made up of five carbon rings, was made using an atomic-force microscope.

From "The Chemical Structure of a Molecule Resolved by Atomic Force Microscopy," by Leo Gross, Fabian Mohn, Nikolaj Moll, Peter Liljeroth, Gerhard Meyer, *Science Magazine,* The American Association for the Advancement of Science, August 1, 2009. Reprinted with permission from AAAS.

The subtleties of what physicists mean when they say that something exists led Steven Weinberg, in his book *Dreams of a Final Theory,* to take a step back and ask, "What after all does it mean to observe anything?" Weinberg analyzed the "discovery" of the electron, usually credited to British physicist J. J. Thomson, for an experiment he performed in 1897. What Thomson actually did was measure the way the "rays" in a cathode ray tube—essentially an old-fashioned TV picture tube—are bent by electric and magnetic fields. He found that the amount of bending was consistent with the rays' being made of tiny particles that carry a definite ratio of charge to mass, and he jumped to the conclusion that these particles actually exist and are the constituents of all forms of electricity, from wires to atoms. But Thomson didn't really see any individual electrons. He did not even really observe the bending of the cathode rays; he simply measured the position of a luminous spot on the tube, downstream from the source of the rays, then inferred the bending, and the charge-to-mass ratio, by employing accepted theory to calculate how the applied fields could have caused the rays to hit the observed luminous spot. And, "very strictly speaking," says Weinberg, he did not even do that: he "experienced certain visual and tactile sensations" that he interpreted as a luminous spot.

As it happens, German physicist Walter Kaufmann performed a very similar (but more precise) experiment at about the same time. But Kaufmann took a different approach to what he considered real. He believed physics should concern itself more strictly with what is observed, so he did not report that he had discovered a new type of particle—the electron. Instead, he reported only that whatever cathode rays are made of, they carry a certain ratio of electric charge and mass. Meanwhile, Thomson went on to perform follow-up experiments, and found that his electron model applied in other venues, such as radioactivity, and when metals are heated. As a result, Thomson is considered the sole "discoverer" of the electron.

In the century following Thomson and Boltzmann, physics changed a lot. Today physicists have little hesitation considering as real objects that have not been observed—and even objects that we believe *cannot* be observed. In particular, consider the quark, a type of elementary particle thought to be inside the proton, neutron, and many other particles. In the early 1960s, Murray Gell-Mann and George Zweig independently invented the concept. Their theory was compelling, and led to new predictions that were confirmed, providing convincing evidence for the quark model. Yet when experimentalists smashed particles together that should have shaken out and isolated the individual quarks, they never found any. Eventually physicists realized why we don't see them: at its root, it is because the attraction between quarks *increases* with distance, as if the quarks were connected by a tight spring.

If the quark picture is reminiscent of Feynman's parton model, which I described in chapter 16, that is because Feynman's partons are indeed the same particles as Gell-Mann and Zweig's quarks. But Feynman used his mathematical model to explain the data from just a single class of experiments, and for that less ambitious purpose he did not have to assume the partons had all the specific (and in one case, peculiar) properties that Gell-Mann and Zweig had postulated for their quarks. At the time, most physicists were unsure whether to consider either the quarks or the partons real, and by using a different name for the particles in his model, Feynman avoided committing himself to endorsing the specifics of theirs. Eventually, however, as Feynman had speculated might happen, the parton model and the quark model—having proved their usefulness not just as "psychological guides" but as a way of producing "other valid expectations"—did "become real" as far as physicists are concerned, even though we have never observed one, and most of us believe we never will.

Einstein stressed the importance of observation, writing, "Pure logical thinking cannot yield us any knowledge of the empirical world; all knowledge of reality starts from experience and ends in it." Today, however, mathematics and observation are more intimately intertwined than ever before. In modern physics, observation cannot be freed from either the human sensory system or the human reasoning system.

As long as they don't conflict, different theories, each valid in the sense that its predictions are confirmed by observation, can present us with different pictures of the reality, all of them valid. As an example of different but nonconflicting realities, in *The Grand Design* Stephen Hawking and I wrote about the worldview of a goldfish in a bowl with curved sides. A freely moving object outside the bowl that an observer there would deem to be traveling along a straight line—as Newton's law requires—would seem to the goldfish to be moving along a curved path. A pioneer goldfish scientist might therefore formulate laws regarding the motion of objects outside the bowl that are different from Newton's laws. Despite that, the goldfish's laws would enable the goldfish to predict correctly what it could see of the motion of objects on the outside, and so the laws would represent a valid picture of reality. Now suppose an exceptionally brilliant goldfish proposed another theory, that Newton's laws apply beyond the boundary of their watery universe, but that light from that outer realm bends as it passes into the water, causing the paths of outside objects to only *appear* curved. This theory offers goldfish scientists a completely different conception of what is going on. Those of us standing outside the fishbowl can argue that the latter is the theory that really describes "reality," but since both theories supply their formulators with the same accurate predictions, each would have to be considered an equally valid picture.

I have argued that today's universe is a result of the laws of physics, that humanity arose from randomness guided by nothing more than the natural selection of evolution, and that our thoughts and feelings are phenomena with origins in the physical brain. It is difficult today, in view of what science knows, to believe in a god that created the universe a few thousand years ago, deposited some creatures in it, and now goes around preventing (or causing) wars, healing (or afflicting) the sick, helping college quarterbacks score touchdowns (or enabling the defense to stop them). But science has taught us that other realities can exist, and that if you look under the hood of everyday life, the workings of nature are very different from those we perceive with our senses. Is there also room for another hidden reality, a reality that includes God?

Even those who understand quantum theory go about their everyday lives employing the model of reality described mathematically by Newton as their working hypothesis. There is nothing to be gained in describing billiard balls by quantum mechanics, or refusing a glass of wine because of uncertainty regarding the momentum of the molecules that compose it. Belief, too, can be a working hypothesis. I once asked a friend whose rationality I respected why she believes in both God and an immortal soul when there is no evidence for either. I expected her to disagree about the absence of evidence, but she didn't. Do your beliefs have to be consistent, she asked? Can you enjoy a film even if you'd be at a loss to describe its merits? Can it speak truth to you even if it is not a cinematic masterpiece? Why is it wrong to believe in a higher power even if you don't have proof? Then she told me of a book published in German, a collection of notes and letters written by people about to be executed for helping Jews survive during World War II. All were written either by people deeply involved in their faith or by children. There was only one exception, she said—a nineteen-year-old secular man who got involved in the resistance movement as a sort of adventure. His letters were different from all the others, she said. He was the only one who feared death.

DEEPAK

If you put a frog in a box and hand it to a scientist, he can tell you many fascinating things about the creature. Luigi Galvani, an Italian physician in Bologna, applied a spark to a frog's leg in 1771 and observed that the leg muscles twitched. Making a connection between electricity and how the body works opened up a new world. It would be fair to say that without Galvani's simple observation, the entire field of neuroscience couldn't exist.

If you take another box and put a human brain in it, scientists once again discover fascinating things, but some essential mysteries cannot be explained—for example, how images are visualized in the cortex, how a brain cell stores memory, and how we come to identify with a self. So from a scientific perspective the brain is a "black box," a system whose inner workings defy explanation. When you put something in a black box, all a scientist can study is what goes in and what comes out. What takes place inside can only be a topic for speculation.

Yet there is a third kind of box that Leonard has been struggling with. In this box you put reality. When you ask a scientist to tell you what's in the box, he runs into huge problems. For example, Leonard grapples with my interpretation of the strange way that atoms, the basic building blocks of the physical world, exist in a shadow realm between the real and the unreal. I rely on the fact that every particle in the universe has its source in "nothing." Naturally, it is very difficult to connect nothing with something, the visible with the invisible. Our back-and-forth has been a tussle over this one problem, really. Leonard ends his last essay by putting science and spirituality in separate compartments, each viewing the universe from its own perspective. I don't think that's good enough, not if we expect a rational Christian, for example, to accept evolution over the book of Genesis. We have to look at the whole picture, subjective and objective. Only then can we stop defending a flawed worldview, whether it's scientific or spiritual. Worldviews are pointless unless they can explain the reality in the box.

Even among the most broad-minded physicists, the mystery of reality borders on the unsolvable. It is both touching and sad to read about the regret that the early quantum pioneers felt when they realized that they had left the physical world—so dependable, reassuring, and available to the five senses—in tatters. Schrödinger, after providing his famous equation that explained the wavelike behavior of particles, wished that he hadn't made his discovery, given the pain and struggle it caused. Einstein refused to accept the strangeness of a world ruled by quantum mechanics. For him, the dismantling of certainty was too unnerving. But there is no doubt that quantum theory was right, so far as calculation goes.

Leonard represents a generation of physicists who have come to terms with quantum reality, but I believe he has paid a high price. In my view, he has finessed the most unsettling facts, even though science is supposedly ruled by facts. The first fact is that all experience occurs in consciousness. This is more than brain processes. The second fact is that if there is a reality outside consciousness, we will never know what it is. Leonard acknowledges that nothing outside the brain can be known, yet he somehow believes at the same time that science is on the right road to all the answers we will ever have. Perhaps the most distinguished physicist to try to account for this discrepancy, Sir Roger Penrose, remained baffled, declaring, "I do not believe that we have yet found the true 'road to reality,' despite the extraordinary progress that has been made over two and half millennia, particularly in the last few centuries. Some fundamentally new insights are certainly needed."

On my side of the debate, these insights have existed for a long time. Reality is pure consciousness. Nothing exists outside it. Its effects are all-embracing. There can be no other answer, yet to arrive at this one, science must drop the illusion that there is a physical world "out there" to which it can cling. Leonard has been clinging with all his might, even as he cites evidence to the contrary.

This reminds me of the fishermen who brave the frozen winter seas off Alaska to catch king crabs. Their job is considered the most dangerous in the world. Their small fishing craft become encrusted in thick ice, making it treacherous to stand on deck, much less carry out the perilous work of hauling up traps heavy with crabs while being tossed about by huge waves.

I can imagine Leonard as the captain, shouting to the first mate to measure the next wave that is about to crash over the bow. The mate raises an instrument to his eye, which tells him it's a thirty-foot swell. "How fast is it coming at us?" the captain shouts, worried that they may capsize. The first mate grabs another instrument, takes a reading, and finds that the wave is approaching at forty knots. But by the time he is ready to shout out the answer, the wave has hit the boat, and it's all that the crew can do to grab the gunwales or the mast, holding on for their lives.

If you substitute a light wave or a stream of electrons for the oncoming ocean wave, this is remarkably like the situation that Einstein and his colleagues found themselves in. Like the first mate, they could take measurements of mass, charge, and spin by stopping physical reality in midprocess and describing whatever they could. Meanwhile, the waves keep crashing over the bow: reality is perpetually on the move, waiting for no one.

Penrose understood how unmanageable reality actually is, saying, "Some readers may still take the view that the road itself may be a mirage. Others might take the view that the very notion of a 'physical reality' with a truly objective nature, independent of how we might choose to look at it, is itself a pipe dream." Leonard seems to feel no ambivalence on these matters. "Shut up and calculate" is always a potential fallback position where hard science is concerned, but reality won't shut up, and the gnawing truth is that our commonsense concept of the physical world has proved a leaky boat indeed.

Let me take a stab at helping any skeptical reader understand why pure consciousness must be the right answer to the question, What is fundamental reality? The thorn in everyone's side when posing this question is that whatever fundamental reality is, it cannot be created. If you plant a stake and say, "This is it. X is the most basic aspect of reality," anybody can raise their hand and say, "But who or what created X?" The creator of X—whether it is God, mathematics, gravity, the curvature of space-time, or any other speculation—must always be more fundamental.

Which means that the source of creation is uncreated—a concept science has an almost impossible time wrapping its head around. Theories about multiple universes don't save us, because even if you hold that there are trillions of other universes, who or what created *them*? One camp speculates that each world creates the other, or that they rise and fall in a cosmic rhythm of birth and death. That doesn't solve the problem, either. Who or what started the rhythm? The uncreated is an intellectual nightmare.

Although we assume that we must be the smartest people who ever lived, the ancient sages of India knew enough to declare that X, the most fundamental reality, has no physical properties. They refused even to give it a name, instead calling it "that" (*tat* in Sanskrit). By calling it pure consciousness instead, I've committed a philosophical sin, making X seem more tangible than it really is. At bottom, I accept the nameless, formless, inconceivable nature of "that."

Here science and spirituality can hold each other in a consoling embrace. Just as the atom disappears once you realize that it has "no physical properties at all" (Heisenberg), the human mind also disappears once you realize that it has no physical properties, either. Atoms emerge from a void that is pure potential; thoughts emerge from a void that is pure consciousness. In the interests of fairness, one must offer a challenge. When you describe the void, you are simply making nonstatements about nonexistence. Aren't you just giving up?

Here we are saved by an unlikely hero, technically known as "qualia," a word that derives from Latin and refers to the subjective aspects of perception. The redness, softness, and scent of a rose are all qualia, for instance, as are the saltiness of salt and the sweetness of sugar. Daniel Siegel has bundled these qualia together with the acronym SIFT, which stands for sensation, image, feeling, and thought. It's a clever acronym because we do sift through the flood of data that bombards us from all sides, transforming it into one or more of these qualia. Going back to Christopher Isherwood's famous phrase, "I am a camera," the rock-bottom reason that you and I aren't a camera—or any other machine—is that a camera doesn't sift through reality, whereas we have no choice but to do so. Peering at the Grand Canyon involves a unique process of sifting, as each of us notices various colors amid the changing light, takes in the smell of the surrounding pines and the sound of the wind rushing up from the canyon floor, and then incorporates all of this into a feeling of awe (or boredom, if your job is to pick up the trash left by tourists), along with personal thoughts as the scene soaks in.

No two people sift the experience of the Grand Canyon in exactly the same way. Two cameras, however, can easily take a pair of photos that are exactly alike. Science pounces on this uniqueness with alacrity, insisting instead that one experimenter must replicate the results of another experimenter for verification. Yet in pretending that a camera is recording reality as it *should* be recorded, science has thrown the sifter out the window. The qualia that have been discarded—those sensations, images, feelings, and thoughts—turn out to be the only things we can really trust. If I send a crab fisherman to sea in Alaska with a sheaf of data about the waves he will encounter, it would be foolish to claim that he is prepared for the hazards ahead. Those huge waves are cold, heavy, pounding, fearsome things—that's their reality. They are nothing *but* qualia.

So the obvious question is, Where do qualia come from?

Neuroscientists claim that they come from the brain. An ancient thinker like Plato claimed that they are part of Nature. Both answers are assumptions. No matter how finely a neuroscientist probes the visual cortex, he will never find the redness of a rose in those squishy gray tissues; he will find only electrochemical soup. No matter how deeply a philosopher looks inside the mind, he will never find the exact point where consciousness suddenly produces velvet redness. The trail ends with an admission that the sensations, images, feelings, and thoughts that constitute reality are irreducible. Qualia rule.

That's why the connection between mind and brain—or to be more general, between mind and anything physical—is called the hard problem. Consciousness won't let you look behind the curtain. Reality is modest; it won't be seen naked. But what if you reverse the hard problem? Instead of asking for a physical explanation of subjective reality, ask for a subjective explanation of the physical world. This tactic works. If you break down a brain cell seeking where the redness of a rose comes from, the cell will eventually vanish into energy waves, and those waves collapse into potentials. If you start instead with redness as an experience, it will again vanish, this time into the silent mind. But when this happens, you won't be left empty-handed. You will still be awake and aware. *That* can't be made to vanish. Moreover, by throwing a mental switch, you can turn silent awareness into the whole physical world. We do it all the time. Even scientists, while claiming that they are being purely objective, are doing it. Consciousness is master of everything that emerges from itself.

Leonard dismisses or ignores arguments that might upset his grip on objectivity. I sympathize. The *Yoga Vasistha,* one of the major texts of Vedanta from India, proposes a startling idea. When describing the ultimate reality, Vasistha says, "It is that which we cannot imagine, but from which imagination springs. It is that which is inconceivable, but from which all thinking springs." To me, this statement is so close to quantum reality that I keep wondering when my scientific friends will jump into the water—and discover that not only is it safe, it's familiar.

There's no terrifying mystery here, nothing to fear. The thing is, we are all in contact with our inconceivable, unimaginable source. As much as Schrödinger and his colleagues rebuked themselves, they got over the pain that came with accepting a quantum world. Now it's time to integrate that world into our everyday working lives, because consciousness is fully capable of embracing both the subjective and the objective aspects of reality. The two don't have to live apart, and in truth they cannot. We are sifting and sifting through every single second of our existence. Many scientists wouldn't trust the inner journey, but I don't trust anyone who has a fixation; materialism is a fixation, which I observe with real sadness. Materialism has caused as much struggle and pain as anything the world has ever witnessed. Our greed for possessions goes hand in hand with our willingness to wage war against those who pose a threat to our possessions, or whose defeat will lead us to own even more. Only in the light of the awareness that binds all human beings together does true safety lie.

Epilogue

LEONARD

In the mid-nineteenth century, a leading physicist in England was asked to assess the "table-turning" phenomenon that had become a craze among people who felt that a type of spiritual contact was occurring during these sessions, allowing them to communicate with the dead. The supposed contact took place as participants sat around a table, resting their hands upon it. After some time, the table would become animated. It would turn, tilt, and move about, sometimes dragging the sitters along with it. Determined to embark on a serious investigation of the phenomenon, Michael Faraday, inventor of the electric motor, one of the founders of electromagnetic theory, and one of the greatest experimental physicists of all time, attended two séances, where he performed a series of technically difficult, intricate, and ingenious experiments that enabled him to understand what was happening. Faraday showed that the motion began as random fidgeting; then at some point the participants' small fidgety movements would coincide and amplify one another until the table moved slightly. The expectant participants followed it, inadvertently amplifying the motion even further until it seemed the table had a mind of its own. The effect was quite dramatic, and the participants, who were unconsciously pulling and pushing the table themselves, not being pulled by it, genuinely believed the motion to be a communication from another realm. But Faraday discovered that it wasn't.

Every once in a while we all come across something mysterious and unexplained. When that happens it is good to be open-minded.

But to passively accept a ready-made answer, without a critical consideration of the alternatives or any serious scrutiny of the "proof," is not being open-minded, it is being empty-minded. Unfortunately, humans seem to be by nature more comfortable with a definite if flimsily supported explanation than with hypotheses requiring more investigation and analysis before the issue can be regarded as settled.

I don't mean to compare Deepak's spirituality, which has its roots in ancient Eastern philosophy and religion, to the nineteenth-century "spiritualist" movement that embraced table moving. I use the example merely to show that science has often, throughout history, examined untraditional ideas. Moreover, sometimes it comes around to accepting them. For example, until Einstein's work in 1905, the idea that measures of space and time are subjective, and depend on the motion of the observer, would have sounded just as foreign and implausible as Deepak's ideas sound to most scientists today. And some of Einstein's contemporaries never did accept relativity. Nevertheless, it soon became mainstream physics. Why? Because relativity's predictions were shown to conform to experimental observation. Unfortunately, Deepak's words and ideas do not.

I have tried in this book to point out where Deepak's arguments clash with what modern science tells us. In response, he has referred to a "stubborn resistance of science to other ways of regarding the cosmos." He argues that scientists are closed to seeing the world other than through their traditional "materialist" lens. Deepak's views of a purposeful universe and the immaterial realm of the mind do not constitute a religion, yet like the religions that address these issues, Deepak's beliefs are far less open to being questioned and altered than are the beliefs of science. The *Catholic Encyclopedia* explicitly warns us that to disbelieve the Christian revelation "involves not merely intellectual error, but also some degree of moral perversity," and that "doubt in regard to the Christian religion is equivalent to its total rejection." Deepak does not go that far, but his key points, too, have traveled to us largely unchanged if not unchallenged since their origination with great Eastern philosophers of centuries and even millennia ago. In science, on the other hand, we are constantly refining our views, and have readily forsaken the orthodoxies of our sages, from Newton to Einstein to Bohr, whenever the evidence required us to do so. Science thrives on doubt. Far more than any religion, science has been open and accepting enough to embrace vast revolutions in its worldview, and seeming heresies on issues like the corruptibility of time and space, and the impossibility of certainty in prediction. Even the materialism that Deepak tells us science holds sacred has been altered as our knowledge of the universe has increased. At first science considered only visible, palpable objects as real; then science grew to accept intangible force fields, unseen atoms, and even unseeable quarks. Science is open to accepting new truths. What it resists is accepting untruths.

Science is open-minded because it has no agenda. Science does not care if the Earth is the center of the universe or just another planet, if the Milky Way is the only galaxy or just one of many billions, or even if our universe is not unique. Science does not take offense at finding that human beings developed from apes or bacteria, that we are gone to dust when we die, or that our consciousness has no magical side to it. Darwin did not approach the issue of the origin of life saying "We must remove purpose from creation." Deepak, in comparison, writes, "If we want to evolve beyond our worst impulses, the only way is through a higher purpose that benefits everyone" and "Spirituality restores purpose and direction to their rightful places at the heart of evolution."

I agree that it is good to lead a purposeful life, but that should not be confused with believing that nature has purpose built into its laws. I also applaud Deepak's vision of how people ought to live and to treat one another. But whereas Deepak and I both would like to see a better world, one in which people have transcended their worst impulses, as a scientist I cannot let the way I *want* the world to be drive my apprehension of the way the world *is*.

One of the issues Deepak feels science is closed-minded about is the existence of a hidden or invisible realm. It is true that historically science has rejected many suggestions of invisible realms. But that's not because science has never examined them. One of the most important traits of a great scientist is curiosity, and over the years scientists from Faraday to Feynman have pondered such issues. But another great trait of the scientist is skepticism, for there is no joy in satisfying curiosity with false explanations. The requirement that our theories correspond to what we observe in the real world has thus far necessitated our always rejecting ideas regarding the immaterial realm.

Events can be deceiving, and discovering their true explanations often isn't easy. The emergence of galaxies, stars, and people from chaos can appear, like tables apparently moving by themselves, to demand some supernatural explanation. When philosophizing one can talk freely about unseen realms, invisible realities, and organizing forces that guide evolution. One can illustrate the ideas with stories and anecdotes, and argue by analogy. One can use everyday language, with its pitfalls of vagueness, and terms with multiple meanings. One can pepper one's prose with satisfying terms like "love" and "purpose." One can even appeal to ancient sages and texts. These arguments may seem attractive. But science answers to a higher authority—the way Nature *actually* works.

When Richard Feynman had the idea to recast quantum theory based on his new interpretation, a reformulation that would give physicists a completely different picture and a new understanding of reality, he too began with simple examples and analogies. But then he spent years making his ideas precise, figuring out all the details, defining exactly what his words and ideas meant, and recalculating almost every quantity anyone had ever calculated using the old formulation in order to check that his form of the theory produced the same predictions—all of which had been confirmed by experiment. Only then did Feynman believe and publish his revolutionary work. For a theoretical physicist to have a new and interesting idea, or even to develop an attractive and plausible new theory, is not uncommon. To have it meet the test of reality and find acceptance is. The scientific approach to truth has brought humanity a wealth of knowledge not attainable by other means.

Deepak has repeatedly brought up the destructive applications of science. But let's not forget that a world that ignores the truth of science is a world left in the darkness of superstition, the misery of ignorance. Centuries ago, the human condition was one of pestilence, filth, hardship, and disease. Think about the improvements in living conditions that have resulted from the scientific revolution. Being a physician himself, Deepak knows that if we relied on his wisdom tradition for our knowledge of the universe, instead of the scientific method, we'd still be falling victim to rampant diseases like smallpox, tuberculosis, polio, and pneumonia, and women would still commonly perish in childbirth; we'd be the victims of dirty and disease-ridden water; and we'd be starving because agriculture could not have kept up with worldwide food demand, nor would reliable contraception methods exist to help people limit the number of children to those they can feed and support. In short, we'd still be dying before middle age because ancient wisdom traditions are no substitute for modern science.

I'm not saying that science has all the answers. Consciousness lies at the heart of Deepak's worldview. It is also science's last frontier. Today science does not even have a good operational definition. We are like Michael Faraday at the start of his career. As he explored what we now call electromagnetism, even the characterization of electricity as positive or negative was controversial. Many analogous debates about the fundamental nature of consciousness take place in science today. We poke around, we make observations, but we are not really sure what it is we are trying to study. Still, there is no reason to believe that consciousness won't be explained. We need not jump the gun and accept that its explanation lies in some unphysical realm.

There are many mysteries in physics today, from the nature of dark matter, to the recent discovery that the expansion of the universe is accelerating, to the possible observations of exotic new types of neutrinos that don't fit into the standard model. Such mysteries could result in a revision of current theories, or in a complete overhaul. Either way, it is natural for scientific theories to keep evolving. When I talk to other scientists about the possibility of identifying a phenomenon that pokes a hole in our current theories, the most common response I hear is a desire for such an anomaly to occur. For while metaphysics is fixed and guided by personal belief and wish fulfillment, science progresses and is inspired by the excitement of discovery. The scientist's dream is to make new discoveries, especially when they mean that established theories must be revised. Scientists discovered two new forces in the twentieth century—the strong and weak nuclear forces—and the same excitement that accompanied those discoveries would reign if we ever found real evidence of another realm of consciousness. All it takes is convincing data to support the idea. If that were to come, many a scientist would enlist in the effort to find more evidence, in order ultimately to prove or disprove the existence of that realm.

I've argued for a worldview grounded in observation and evidence, and I've argued that such a viewpoint need not deny the richness of the human spirit, or the wonder of the universe. As Einstein wrote, concerning the idea that human behavior is governed by nothing more than the laws of nature, "This is my belief, although I know well that it is not fully demonstrable. [But if] one thinks out to the very last consequence what one exactly knows and understands, there will be hardly any human being who will be impervious to this view, provided his self-love does not ruffle up against it."

Admittedly, our self-love makes it difficult to accept a worldview in which human beings do not play a central role in the universe. But science's ultimate triumph lies in the integrity of its methods, the openness of its point of view, the eagerness of its embrace of the truth. Science may never have all the answers, but it will never stop looking for them, and it will never take the easy way as it continues on its search for understanding.

DEEPAK

To many readers, there is no war of two worldviews, or if there is, one combatant is puny and unarmed while the other possesses tanks, robot drones, and smart bombs. Science is fully armed, while a new spirituality, divorced from religious dogma, is a fledgling. I'd suggest that the war doesn't need to be fought anymore, because it's already over. Hidebound science is ready to topple, making way for a new paradigm where consciousness takes center stage. Don't expect the bodies of fallen physicists littering the field. The outcome won't be the vanquishing of science but its expansion. The expanded version will be able to admit into evidence something that Leonard shuns: a purposeful universe. (When Leonard says that I am clinging to precepts from thousands of years ago, he can't be serious, given how much up-to-date science the new spirituality has come to terms with.)

He himself hits on the guiding principle of an expanded science, which "answers to a higher authority—the way Nature *actually* works." Unfortunately, he hasn't been able to follow his own prescription. Faced with evidence about post-Darwinian evolution, the quantum basis of consciousness, and the futility of equating the brain with the mind, Leonard runs for shelter in cherished beliefs that forward-looking science is abandoning with greater and greater speed. I invite him to jump in the water—it's not scary—but like the *Catholic Encyclopedia* he somewhat bizarrely cites, he has deeper concerns (scientific salvation, perhaps?) that make it forbidden to accept a spirituality that is consistent with science. Anyone allied to rooting mind in matter will continue to ignore the anomalies that crack their worldview.

Leonard is in favor of leading a purposeful life, only he wants to divorce it from science. I've always been struck by the way scientists wed themselves to the dogma of a random universe, one totally devoid of meaning, when it's obvious that every moment of life embraces the things that matter to us, even if your goal is as small as making it through the day, finishing a mystery novel, or picking up the kids after soccer practice. If our life has meaning, it must have come from somewhere.

For me to declare that the war is over, I must offer evidence. These essays have indicated numerous trails of evidence—from the plasticity of the brain to the fluidity of genes, from the quantum vacuum to the domain outside space and time—to meet the call for "new insights" that Sir Roger Penrose has sounded. Twenty-five years ago, my medical colleagues in Boston refused to believe that there was a mind-body connection. Now it's accepted without question that our thoughts, feelings, and moods are conveyed instantly to every cell in the body. The cell membrane receives news of the world, inner and outer, and on a microscopic level it *is* the world, written in molecules. Back then, when a professor of medicine smirked at the notion of the mind affecting the body, I would blurt out, "How do you wiggle your toes? Isn't your mind sending an order to your feet?"

I've declared repeatedly that I am not defending any conventional God. But spirituality cannot be artificially segregated from the essence of religion. Both depend on a personal journey, leading in the end to the transformation of consciousness. The invitation to begin such a journey comes from reality itself. I firmly believe that reality wants to be known, and human evolution answers to that call. Science is one answer, but it can't hog the road; spirituality is just as valid an answer.

Science shouldn't be the enemy of the inner journey, and I feel disheartened if Leonard believes that his view of a "higher authority" forbids inner exploration, as if table rapping in a Victorian séance should be our model for spirituality. Does anyone think that the Buddha and

Plato ran séances? Yet there's no reason to make rhetorical hay here. The world's great spiritual teachers were Einsteins of consciousness. They provided principles and discoveries fully as valid as those of Einstein, who had his religious doubts but never lost sight of the awe and wonder that he felt were essential to all great scientific discoveries.

Leonard places great store in doubt as a tool of science. I can only agree, but a rigid, hostile skepticism does no one any favors. Skeptics squat by the road like guardians of truth, letting no one pass who doesn't come up to scratch. They never realize that they can see only what their paradigm tells them to look for. If you judge a person only by how well he plays pool, Mozart won't pass scrutiny, but the fault is in your lens.

I was once talking on mind and body before an audience in England when a loud, red-faced man jumped to his feet shouting, "This is all garbage. Don't listen to him. It's crap!"

The audience stirred uneasily, and I was a bit shaken. "Who are you, sir?" I asked.

"I'm the head of the UK Skeptical Society," he replied.

"I doubt it," I said, and the audience burst out laughing.

Leonard comes close to joining the Society for the Suppression of Curiosity, which is where blanket skepticism leads. But I imagine he is as guided by awe and wonder as Einstein, so let me speak to those qualities. At the instant of the Big Bang, the laws of nature apparently came about within 10^{-43} seconds—an unimaginably short blink in which to assemble every ingredient of the known universe inside a space trillions of times smaller than the period at the end of this sentence. Nothing existed during the "quantum epoch" that preceded this instant except for a sea of roiling energy. Even that is conceptually shaky, because there were no physical laws, and therefore nothing like electromagnetism existed, either.

The human brain, if you believe in strict materialism, was also predetermined in this roiling energy soup billions of years ago. If so, then we are the product of what came next: this astonishingly fine-tuned universe, where dozens of constants are perfectly meshed in such a way that a change of one part in a billion would have defeated the whole venture. You are able to read and think—along with playing at billiards or the game of love—only because of what came after 10^{-43} seconds. Without light, gravity, and electrons, not to mention time and space, none of us would be here. What came before is unknowable, and for that reason alone, science is reduced to conjectures no less fanciful than what I have been proposing. When we argue over where the cosmos came from, the playing field gets flatter every day.

In fact, fanciful is being kind. Materialism cannot venture anywhere before the creation of matter. Objectivity cannot venture anywhere before there were objects to observe. If the fate of the universe was decided in a single moment, why can't it be a creative moment? Leonard's thundering "no!" makes little sense. It's not as if *his* method will get us anywhere. Our subjectivity connects us to the primordial impulse to make something out of nothing; otherwise, we deprive ourselves of creativity, deep intelligence, and free will.

Ordinary people aren't going to give up emotions and inspiration just because science sniffs at subjectivity. Science shouldn't be so edgy and defensive. Vandals aren't going to smash their way into laboratories and throw Bibles at the equipment. Despite reactionary religious activity on the fringes, we all accept that science represents something enormously good and progressive. The ivory tower would be a modern replacement for the sacred city upon a hill, but unfortunately, from that tower rained down not just good things but the atom bomb, biochemical weapons, and nerve gas.

Most scientists wince at the existence of weapons research and then go on about their business. The rise of diabolical creativity seems unstoppable. Other scientists join the profitable enterprise of death with relish. One must be decisive here: a world ruled completely by science would be hell on earth. Being wedded to rational thought is acceptable inside the lab, but once science ventures to dismantle faith, striving, love, free will, imagination, emotion, and the higher self as so many illusions cooked up in our fallible brain, a rescue effort must be mounted, and quickly.

I don't mean to embarrass anyone by my fervor—we all know the destructive power of fervor when it's attached to religious intolerance. But the time is growing late. Millions of people have abandoned organized religion. Almost a hundred years ago Freud derided religious faith as a rearguard action in defense of the indefensible. But aspiration is defensible, and it can't be fulfilled by science—not unless science is willing to break down the walls that falsely separate the inner and outer worlds. Ten years ago it was considered unthinkable to be interested in consciousness and still preserve a respectable scientific career. Today one can attend conferences where hundreds of scientists across every field present panels on consciousness, and the word "quantum" is tossed around to describe brain processes, photosynthesis, bird migration, and cell formation. Right under the nose of physics, brilliant minds are creating a new field, quantum biology.

Which means that to have a vision of a new, expanded science is no longer a folly. Clearly the rescue operation needs to expand much wider, however. All around us people ache with emptiness and yearning; there's a vacuum to be filled, and it's a spiritual vacuum. What other word really fits? Only when people are given hope that this ache can be healed will we truly know what the future holds. Let science join in the cure, because otherwise, we may wind up with marvels of technology serving empty hearts and abandoned souls.

Acknowledgments

DEEPAK CHOPRA

Nothing is more gratifying to a writer than discovering that his books have become a family affair. In this case, the family extends to a staff at the Chopra Center who is tireless in keeping every detail in place and on schedule. Warmest thanks to Felicia Rangel, Tori Bruce, and the most indispensable of all, Carolyn Rangel. I can't imagine getting more sincere and understanding support than I do from my publishing team, including Julia Pastore, Tina Constable, Tara Gilbride, and Kira Walton. No book can come out right without a patient, talented editor, and Peter Guzzardi, who has traveled the long road with me for many years, proved once again that he is one of the best in the business. In Leonard, I found a stimulating and generous mind and a man who quickly became a warm friend.

This extended family begins at home, with my wife, children, and grandchildren. That's also where every fulfillment leads, and no amount of thanks could be enough.

LEONARD MLODINOW

Deepak and I have different worldviews, but one thing we agree on is our gratitude to our publishing team, especially Julia Pastore, Tina Constable, Tara Gilbride, and Kira Walton; our editor Peter Guzzardi; and Carolyn Rangel, who works for Deepak, but was indispensable even to me. And to Deepak, gracious even while we were at times slugging it out. In addition I would like to thank Beth Rashbaum for her insightful comments on the manuscript. I am also grateful to the many others who read all or parts of various drafts, and shared their input—Donna Scott, Markus Poessel, Peter Graham, Mark Hillery, Christof Koch, Ralph Adolphs, Keith Augustine, Michael Hill, Uri Maoz, Patricia Mindorff, and the "Mavericks"—Martin Smith, Richard Cheverton, Catherine Keefe, and Patricia McFall. And of course, my wonderful agent, Susan Ginsburg, advocate, cheerleader, critic. Finally, also, thanks to my family, who had to endure both my long hours away, and my engaging them in endless obsessive book-related conversations that would have made anyone *but* my family wish I would go away.

Index

addiction, 190, 195
Adolphs, Ralph, 203
Afar Triangle, 123
aging, 146
agouti mice, 140
Aham Brahmasmi, 41
Ahmadinejad, Mahmoud, 276
Alekhine, Alexander, 223
al-Ghazali, Abu Hamid, 198
altruism, 128, 155–56, 157, 161, 165–68
amino acids, 138
animals:
  behavior in, 174, 201–2, 208
  communication in, 128
  genetic similarity in, 166
  intelligence in, 105
  internal clocks in, 67
  medieval treatment of, 126
  self-awareness in, 47
  social order among, 127, 161
  time perception in, 75–76
  see also humans; primates; *specific animals*
anthropocentrism, 123
anthropomorphism, 85
ants, 165
Ardi, 123
Aristotle, worldview of, 175–76
arsenic, 97
artificial intelligence, 192, 205–25
  meaning in, 217–19, 222, 223
  measurement of, 210–13
Asimov, Isaac, 217
atheism, 153, 184
atoms, 53–54, 80, 98, 286, 289
  modeling of, 280–81
  time measurement with, 68, 70, 71
autism, 129
autopoiesis, 82–83, 148
  see also growth
Avery, Oswald, 138

Bach-y-Rita, George, 197
Bach-y-Rita, Paul, 196–97
Bach-y-Rita, Pedro, 196–97
balance, sense of, 196
beauty, 130–31, 182
behavior, animal, 174, 201–2, 208
behavior, human, 189–204
  altruism in, 161, 168
  brain and, 189–204, 239
  genetics and, 147, 148, 202
  influences on, 191, 203–4
  see also free will
belief:
  scientific view of, 253–55, 256–57, 268–76, 278–79, 284–85
  spiritual view of, 260–67
  see also God

Believing Brain, The (Shermer), 255
Bhagavad Gita, 40–41, 267
Bible, 276
Big Bang, 4, 24, 30, 32, 52–54, 302
  material created by, 36–37, 53–54
  preexisting conditions of, 41, 43–44, 52, 73, 75, 81
  see also universe, emergence of
Big Bang theory, 26–28, 29, 30, 253
biology:
  evolution in, 57, 59, 111, 113
  life in, 96–97
  mind evidence in, 178
  molecular, 136
  quantum, 304
  time perception in, 68
bisphenol A, 140
Blackburn, Elizabeth, 76
Blake, William, 72
blindness, 195–96
blindsight, 48
Blind Watchmaker, The (Dawkins), 58
Bohm, David, 55
Bohr, Niels, 55–56, 246, 271
Boltzmann, Ludwig, 280
Boole, George, 206
brain, human, xviii, 40, 171–241
  and behavior, 189–204, 239
  complexity of, 16, 142, 207
  as computer, 205–25
  conscious experience in, 73, 125, 255–56
  damage to, 177–78, 184, 185, 191, 195–97, 202–3, 215, 240, 278–79
  determinism and, 124, 190–91, 192, 193, 194, 202
  evolution of, 129, 157, 158, 302
  fear processing in, 194–95
  feedback loops in, 204
  information processing in, 214, 215
  language function in, 178
  mastery of, 184–85, 193–94, 195, 197, 204
  mathematical ability in, 280
  and morality, 128, 131, 202–3
  neuroplasticity in, 195–96, 215, 224
  perception in, 68, 130–33, 199–200, 212, 231, 232, 237–39, 247, 279
  physical stimulation of, 177, 193, 255
  pineal gland in, 176
  scientific understanding of, 174, 176–78, 204, 214–15, 239, 284, 286
  time experience in, 67, 73, 75, 76
BrainPort, 196
Brownian motion, 281
Buddha, 4, 5, 34, 45, 73, 250
Buddhist monks, 73, 183–84, 204
Bunsen, Robert, 268
Bush, George W., 276

cancer, 145, 146
Carroll, Sean, 116
catalysts, 88
Catholic Church, 261
Catholic Encyclopedia, 295
cells, 17, 181
  division of, 80, 81, 82, 83
  life spans of, 156
Chalmers, David, 174, 175
chemicals, organic, 54
chess, 208–9, 210, 223
children, 105, 128–29, 203, 263
chromosomes, 139
circadian rhythm, 67
Clarke, Arthur C., 109
clownfish, 153, 155
coma, 185, 192
Comings, David, 255
compassion, 184
computational theory of mind, 214, 236
computers, 206–7, 208–10, 216–17, 223
  see also artificial intelligence
computer viruses, 97
Comte, Auguste, 268
consciousness:
  evolution of, 159, 160, 246–48, 249–51, 260, 263–67, 301
  human understanding of, 7–8
  pure, 74, 75, 264–67, 288, 289

spirituality's emphasis on, xviii, 4, 6, 7, 10, 221, 225
consciousness, human, 9, 120, 171–241
free will and, 191
scientific understanding of, 16–18, 47–48, 174, 181, 199, 298, 304
time and, 72, 73–74
see also brain, human; mind, human
consciousness, universal, 39–50, 79–90, 227–41
early scientific beliefs about, 48, 95–96
scientific view of, 40, 41, 42, 46–50, 52, 62, 86, 87, 131, 186, 234–41
spiritual view of, 35–36, 37, 40–45, 47, 49, 52, 55–56, 81–84, 87, 89, 96, 104, 119–20, 146, 149, 155–56, 185–86, 228–33, 235, 237, 255, 287, 289, 291, 298
corpus callosum, 177
cosmos, cosmology, xviii, 21–90
expansion of, 26, 52–53, 59
flatness problem in, 27–28, 29
horizon problem in, 27, 29
inflation in, 28–30
life cycle of, 35, 59–60, 118
matter in, 53–54
"nothingness" in, 31, 32, 33–34, 286, 302
physical characteristics of, 26–27, 30
scientific understanding of, 16, 24–31
vacuum fluctuations in, 30–31
see also Big Bang theory; universe
creationism, 108–9, 117, 119, 162
creation stories, 4, 24, 32, 46–47, 117
spiritual basis for, 33, 34–36, 37
see also Big Bang
creativity, 35, 84, 102, 104, 120, 219, 222, 303
Crick, Francis, 99, 136, 138
culture, 127, 168–69

Darwin, Charles, 46, 110, 111, 122, 123, 144, 152, 153, 159, 160, 161, 164, 296
see also evolution, theory of
Dawkins, Richard, 58, 165, 166, 255–56
death, 81, 100, 104, 145, 255
Deep Blue, 208–9, 223
Descartes, René, 175, 176, 229
Descent of Man, The (Darwin), 46
Dialogues Concerning Natural Religion (Hume), 109
DNA, 142–43, 155, 156, 164, 167
adaptation of, 97
complexity of, 36
discovery of, 137–38
emergence of, 35, 41, 54, 58, 87–88, 104, 110, 120
environment and, 148, 149
genes in, 138
importance of, 106
noncoding, 141, 149
self-replication of, 81, 82, 136, 145
structure of, 99, 136, 138
see also genetics
Doctors Without Borders, 228
dogs, intelligence in, 105–6
Dreams of a Final Theory (Weinberg), 281
Drunkard's Walk, The (Mlodinow), 60
Dyson, Freeman, 36, 104, 185

Earth:
composition of, 53, 54, 61, 103, 113
ecology of, 156
entropy on, 116
life on, 35, 54, 55, 61, 80, 82, 142–43
shape of, 162, 164
Einstein, Albert, 11, 24, 33, 34, 37, 69, 70, 101, 136, 165, 219, 257, 272, 279, 281, 283, 287, 295
electrons, discovery of, 281–82
elephants, 111–12
ELIZA, 211, 217
embryos, 83, 148
emergent properties, 98
emotion:
and belief, 271–72
neurological basis for, 174, 177, 181, 284
spiritual view of, 191–92, 194–95

entropy, 35, 113–16, 118, 180
environment:
  and genetics, 137, 140–41, 145, 147–49, 166
  and natural selection, 111, 112
enzymes, 88
epigenetics, 137, 140, 141, 149
epilepsy, 177
equilibrium, 97, 99, 100, 114–15
Ethics (Spinoza), 252–53
evolution, 101
  in biology, 57, 59, 111, 113
  cooperation in, 155, 156–57, 164, 165–67
  cultural, 168–69
  definition of, 57
  directed, 111
  feedback loops in, 158–59
  and genetics, 154
  of human brain, 129, 157, 158, 302
  of human mind, 219–20, 223
  inelegant design in, 112, 141
  instinct and, 270–71
  in physics, 113–16
  progression of, 87–88
  soft inheritance in, 137, 147
  spiritual view of, 35, 52–56, 57–58, 103, 104, 152–60, 161, 164, 165, 219–20, 223
  of universe, 51–63, 86, 103
evolution, theory of, 151–69
  acceptance of, 4, 122, 152, 287
  adaptation in, 153
  application of, 58–59
  limitations of, 159–60
  opposition to, 18, 46, 110–11, 117, 119, 253
  precursors to, 137, 141
  revisions of, 156–57, 161–62
  survival of the fittest in, 152, 154
  see also natural selection
Exodus, 126

Fabre, Jean-Henri, 208
Faraday, Michael, 294
fatty acids, 88
fear, 194–95
Feynman, Richard, 19, 235–36, 256–57, 275, 280, 283, 297
fMRI, 183
Franklin, Rosalind, 99
free will, 130, 146, 154, 190, 191, 193, 194, 195
Freud, Sigmund, 304
fruit flies, 174, 201

Galilei, Galileo, 68, 280
Gallup poll, 247, 276
Galvani, Luigi, 286
Gamow, George, 271
Gazzaniga, Michael, 206
Gell-Mann, Murray, 282, 283
general relativity, 24–31, 34, 136
Genesis, 32, 100, 117, 287
genetics, 98–99, 104–5, 124–25, 135–49, 164
  altruism and, 155–56, 157, 165–68
  and animal behavior, 201
  determinism and, 145, 146–47, 190
  environmental effects on, 137, 140–41, 145, 147–49, 166
  and evolution, 154
  and human behavior, 147, 148, 202
  spiritual view of, 144–49
  see also DNA
genome, human, 139–40, 141, 144, 156
Gilbert, Daniel, 215
God, xviii, 119, 198, 240, 243–92
  American belief in, 276
  Aquinas's argument for, 89–90
  scientific view of, 252–57, 275–76
  spirituality and, 4, 6, 7, 37, 55, 56, 246–51, 260–67, 301
  see also belief
Gödel, Kurt, 220, 221
Gödel's incompleteness theorem, 220–21, 222
Goodbye to Berlin (Isherwood), 230
Gould, Stephen Jay, 168
Grand Design, The (Hawking and Mlodinow), 90, 283

gravity, 25, 26, 30
law of, 163–64, 272
Greeks, ancient, xvii, 182, 220
Greene, Joshua, 202
growth, 86, 96, 156
Guth, Alan, 28, 29

hadrons, 256–57
Hamlet, 36, 122
Harris, Sidney, 256
Hawking, Stephen, 33, 40, 42, 46, 50, 90, 283
Heider, Fritz, 85
Heisenberg, Werner, 246, 289
Helmont, Jan Baptist van, 94
hemoglobin, 54
heredity, 58, 137, 138–39, 144
Himalayan rabbits, 140
Hitler, Adolf, 230
Hofstadter, Douglas, 208
Holocaust, 62, 234
homeostasis, 96–97, 99
hominids, 123, 127, 157
Homo erectus, 127, 157–58
Homo habilis, 127
honeybees, 155, 156, 165–67
Hubble, Edwin, 26
humanity, 16, 121–33
physical evidence of, 123, 124
science and, 12–13, 17, 18–19, 49–50, 62–63, 90, 133, 236, 241
scientific view of, 126–33, 240–41, 284
social order of, 126–29, 161
spiritual view of, 122–25
unpredictability of, 124–25, 129, 145
see also brain, human; mind, human
humans:
cultural evolution of, 168–69
genome of, 139–40, 141, 144, 156
internal clocks in, 66–68
irrationality in, 219
physical order in, 97, 99, 100, 114, 115, 156
spiritual evolution of, 153, 154, 159–60, 161
see also brain, human; consciousness, human; DNA; mind, human
Hume, David, 109, 115
Huxley, T. H., 46
hypothalamus, 67

I, Robot (Asimov), 217
IBM, 209
Incoherence of the Philosophers, The (al-Ghazali), 198
Indian spirituality, 5, 10, 41, 45, 74, 76, 124, 249, 266–67, 289, 291
intelligence, 104, 105–6
intelligent design, 108–9, 117, 119
International System of Units, 68
Isha Upanishad, 8
Isherwood, Christopher, 230, 290

James, William, 250, 271–72, 275
Jeans, James, 42
Jeopardy, 209, 210
Jesus Christ, 4, 5, 14, 122, 182, 250
Jung, Carl, 4

Kaminski, Juliane, 105
Kant, Immanuel, 279
Kasparov, Garry, 208, 209, 223
Kaufmann, Walter, 282
Kekulé, Friedrich August, 219
Kepler, Johannes, 48
kin selection, 168
Kirchhoff, Gustav, 268
Koch, Christof, 177, 178, 181, 212
Koch-Tononi test, 212–13
Krishna, 40, 44, 50
Krishnamurti, J., 246

Lamarck, Jean-Baptiste, 137, 141
language, 128, 178
Lao-tzu, 4
Laplace, Pierre-Simon, 269
Large Hadron Collider, 18
Led Zeppelin, 131–32
Lemaître, Georges, 26

life, xviii, 91–169
biology's definition of, 96–97
complexity of, 142, 157
definition of, 83–84, 85–87, 93–106
on Earth, 35, 54, 55, 61, 80, 82, 142–43
emergence of, 88–89, 103–4, 113, 122
historical views of, 94–96, 123
physics' definition of, 98–99
renewal of, 102, 262–63
scientific criteria for, 85–87, 96–99
scientific view of, 94–100, 104, 106, 108–16, 125, 126–33, 136–43, 146, 161–69
spiritual criteria for, 83–84, 104
spiritual view of, 101–6, 117–20, 122–25, 144–49, 152–60, 262–63
of universe, 79–90, 95–96, 101–2, 106
see also DNA; evolution, theory of; humanity
life cycle, 84, 86
light, bending of, 25
Lloyd, Seth, 236
Loewi, Otto, 219
love, 182, 192, 236, 240
Lucy, 123, 157

McCullers, Carson, 207
maggots, 94–95
manifestation, 84
Marshall, Barry, 15
materialism, 6, 9, 13–14, 33, 36, 56, 101, 103, 104, 106, 118, 119, 146, 221, 261, 292, 296, 303
mathematics, 252, 253
belief and, 268
and human brain, 280
importance of, 34, 42, 49
limitations of, 32, 34, 36
logical systems in, 220–21, 222
observation and, 283
physical laws' embodiment in, 108
probability in, 60–61
unprovable assumptions in, 221, 222
Maxwell, James Clerk, 179
Maxwell's Demon, 179–80, 186
meditation, 76, 184, 193, 232, 261, 263
memory, 66–67, 183, 191
Mendel, Gregor, 137
Mercury, 25
metabolism, 96, 99
Miescher, Friedrich, 137–38
Miller, William, 60–61
mind, human, 171–241
beliefs in, 253–55, 256–57, 269–70, 278–79
evolution of, 219–20, 223
historical theories about, 175–76
illusion of control in, 269–70
naive realism in, 279
observation in, 229, 232, 283
perception in, 230–32, 290–91
scientific understanding of, 174, 176, 178–81, 182, 183, 187, 199, 215, 229, 232–33, 236, 237, 239–40, 254, 301
spiritual view of, 35, 101, 123, 182–87, 190–97, 198–99, 216–25, 228–33, 237, 239, 255, 267, 291, 295
see also behavior, human; brain, human; consciousness, human; emotion; intelligence; sensation; thought
mind-body dualism, 175, 176
Mindsight (Siegel), 229, 239
Minsky, Marvin, 9
Mlodinow, Nicolai, 234
Mlodinow, Olivia, 90
molecules, 54, 55, 80, 88, 98, 99, 111, *281*
morality, 5, 13, 128, 129, 131, 202–3
mystical experiences, 184

naked mole rats, 127
Namath, Joe, 74
Native Americans, worldview of, xvii
natural selection, 57, 58, 110, 111–12, 141, 162, 164, 284

Neanderthals, 127, 154
Needham, John, 95
nematodes, 174–75
neurons, 16, 207
neuroscience, 98, 130, 286
consciousness experiments in, 17
materialistic view in, 9, 125, 182, 191, 192–93, 194, 195, 199, 204, 214–15, 291
technological advances in, 183
time perception in, 68
see also brain, human
Newton, Isaac, 12, 24, 162, 163
Newtonian physics, 25–26
Nietzsche, Friedrich, 46
nucleotides, 138

Old Testament, xvii, 225
On the Origin of Species (Darwin), 46
optical illusions, 230–31, 237–39
orderliness, 34–35, 104
Ornish, Dean, 147
oxytocin, 192, 201–2

Paley, William, 109
paramecia, 82
partons, 256–57, 283
Pascal, Blaise, 219
Pasteur, Louis, 96
Pauling, Linus, 137
Pavlovian conditioning, 124, 125
Penfield, Wilder, 193
penguins, 154
Penrose, Roger, 42, 44, 287, 288, 301
perception:
human, 68, 130–33, 199–200, 212, 230–32, 237–39, 247, 279, 290–91
qualia of, 290–91
of time, 66–68, 69, 71, 72, 74, 76–77
phantom limbs, 223–24
phosphorus, 97
photons, 116
physics, 252
evolution in, 113–16
faith and, 272–75
frontiers of, 9, 32
laws of, 108, 117–18, 130, 131, 133, 178–79, 180
life in, 98–99
mental models in, 256–57, 280–83
mind and, 178–81
mysteries in, 298
Newtonian, 25–26, 130, 163–64
particle, 31, 43
reality in, 279–83, 287
standard model in, 272–75
time measurement in, 68–71
wave function collapse in, 43
see also cosmos, cosmology; general relativity; quantum theory; science
Planck scale, 42
Plato, xvii, 175, 291
Platonic values, 7, 42
Pluto, 130
prefrontal cortex, 184, 190–91, 203
primates, 13, 105, 129
proteins, 88, 97, 138, 139, 140
protons, 60, 61
pseudogenes, 141
psychology, 130, 194, 279
psychopaths, 203
psychotherapy, 217

quantum theory, 29–31, 32, 40, 98, 129, 284, 287, 297
spiritual view of, 32, 33, 34, 35, 40, 45, 74–75, 185, 186, 187, 287, 291–92, 302
quarks, 282–83

Ramachandran, Vilayanur, 223–24
reality, 262, 277–92
in physics, 279–83, 287
scientific view of, 278–85
spiritual view of, 286–92, 301
see also consciousness; quantum theory; universe
Redi, Francesco, 95–96
reductionism, 55, 62, 118, 123, 125, 187

relativity, 70–71, 73, 74
  see also general relativity
religion, 80, 221, 301
  abandonment of, 249, 260–61, 304
  failings of, 4, 5, 6, 8, 13, 52, 159, 161, 248, 295
  science's clash with, 4, 5, 8–9, 18, 32, 52, 119, 120, 123, 152, 276, 303
Repo Man, 212
reproduction, spontaneous, 84
rocks, 103
Rogers, Carl, 217
Romans, ancient, worldview of, xvii
Ruelle, David, 280
Rumi, 45, 186
Russell, Bertrand, 240, 269

samadhi, 74, 75
Schrödinger, Erwin, 98, 99, 122, 287
science, 252, 253
  belief and, 268–69, 284–85
  definition of terms in, 15, 57, 236
  dependency on, 253, 298
  failings of, 5–6, 9, 15, 101, 298, 303
  humanity and, 12–13, 17, 18–19, 49–50, 62–63, 90, 133, 236, 241
  and morality, 5, 13
  observation in, 283–84
  religion's clash with, 4, 5, 8–9, 18, 32, 52, 119, 120, 123, 152, 276, 303
  worldview of, xviii, 6, 8, 9, 11–19, 295, 296, 298–99
scientific method, 14–16, 28, 49, 62, 90, 163–64, 236, 253, 295, 297
séances, 294
Seckel, Al, 131–32
self-awareness, 47, 55
selfish gene, 155–56, 157, 165
self-knowledge, 101
sensation, 174, 181, 196, 223–24, 267
sheep, maternal behavior in, 201
Shermer, Michael, 255
Siegel, Daniel, 229, 232, 239, 240, 290
Simmel, Marianne, 85
sin, 152
sleep, 47, 73
Socrates, xvii
somatophrenia, 278–79
soul, 267
Spallanzani, Lazzaro, 95
spectroscopy, 268
Spinoza, Baruch, 252–53
spiritual hypothesis, 6–7
spirituality:
  consciousness's importance to, xviii, 4, 6, 7, 10
  disciplines in, 193–94, 204, 263, 266
  failings of, 13, 16, 18
  and God, 4, 6, 7, 37, 55, 56, 246–51, 260–67, 301
  human impulse toward, 12
  personal satisfaction and, 260
  principles of, 34–36, 37–38, 54
  scientific objections to, 32–33
  teachings of, 4, 5, 6–7, 302
  worldview of, xviii, 4–10, 13, 14, 16, 18, 19
  see also religion
split-brain patients, 177–78
spontaneous generation, 94, 95, 96
Stannard, Russell, 182
stars, 53, 59, 268
stimulus response, 96, 99
string theory, 30
Stumbling on Happiness (Gilbert), 215
Sturtevant, Alfred, 174
sun, 53, 116, 269
supernovas, 53, 59, 61
synchrony, 69

Tagore, Rabindranath, 149
telomerase, 76
telomeres, 146
Theory of Everything, 123
theory of mind, 129
thermodynamics, second law of, 115, 180
Thomas, Gospel of, 122, 125
Thomas Aquinas, Saint, 89
Thomson, J. J., 281, 282
Thomson, William, 179
Thoreau, Henry David, 248

thought, 75, 194, 228
neurological basis for, 177, 178, 181, 239, 284
spiritual view of, 103, 178–79, 182, 289
time, 65–77, 146
Tononi, Giulio, 212
transcendence, 250–51, 263–67
tubeworms, 155
Turing, Alan, 210, 214
Turing machine, 214
Turing test, 210–12, 213
twins, 147, 166
2001: A Space Odyssey, 217

universe, xviii, 21–90
Aristotle's view of, 175–76
boundaries in, 228–29
death in, 81, 100, 104
design in, 107–20
ekpyrotic, 86
emergence of, 23–38, 41, 43–44, 81–84; *see also* Big Bang; creation stories
entropy in, 35, 113–16, 118
evolution of, 51–63, 86, 103
forces in, 102
historical concepts of, 11, 109–10, 119, 275–76
life of, 79–90, 95–96, 101–2, 106
potential in, 42–45, 75, 83, 289
randomness in, 34–35, 36, 37, 41, 52, 54, 55, 57–58, 60–61, 62, 112, 117, 119–20, 160, 162, 284, 301
renewal in, 80, 102, 262–63
scientific view of, xviii, 6, 8, 9, 12, 13–14, 19, 24–31, 32–33, 34, 36, 37, 40, 41, 44, 45, 52, 54, 55, 57–63, 80, 82, 85–90, 103, 108–16, 119, 129–30, 229, 256, 257, 261, 284, 286–87, 288–89, 295, 296–97, 300–301
self-awareness in, 55
spiritual view of, xviii, 4, 8, 9–10, 13, 14, 18, 19, 32–38, 42–45, 52–56, 60, 80–84, 85–87, 89, 101–2, 106, 117–20, 228–29, 262–63, 275, 286, 289–90, 291–92, 295, 301, 302–3
unpredictability in, 129–30, 187
see also consciousness, universal; cosmos, cosmology
universes, multiple, 40, 289

vasopressin, 202
ventromedial prefrontal cortex (VMPC), 202–3
viruses, 97
vitalism, 96
voles, 201–2

Wadhwa, Pathik, 148
Walden (Thoreau), 248–49
war, 128
Warren, Robin, 15
wasps, 208
Watson (*Jeopardy*-playing computer), 209
Watson, James, 99, 136, 138
Weinberg, Steven, 18, 281, 282
What Is Life? (Schrödinger), 98–99
Wheeler, John, 43
wholeness, 34, 83, 86, 106, 265
Wilberforce, Samuel, 46
worldviews:
clashes of, xvii–xviii, 1–19
scientific, xviii, 6, 8, 9, 11–19, 295, 296, 298–99
spiritual, xviii, 4–10, 13, 14, 16, 18, 19

Yoga Vasistha, 291

Zoroaster, 249
Zweig, George, 282, 283

About the Authors

DEEPAK CHOPRA is an internist/endocrinologist by training and the author of more than sixty books translated into more than eighty-five languages, including numerous *New York Times* bestsellers in both the fiction and nonfiction categories. He is recognized as a leading figure in the field of emerging spirituality.

www.DeepakChopra.com

LEONARD MLODINOW received his doctorate in theoretical physics from the University of California at Berkeley. He teaches at Caltech and is the *New York Times* bestselling author of *The Drunkard's Walk: How Randomness Rules Our Lives* and the #1 *New York Times* bestseller *The Grand Design,* which he coauthored with Stephen Hawking. His other books include *Euclid's Window: The Story of Geometry from Parallel Lines to Hyperspace* and *Feynman's Rainbow: A Search for Beauty in Physics and in Life.* He also wrote for *Star Trek: The Next Generation.*

www.its.caltech.edu/~len